SCIENTIFIC DEVELOPMENT AND MISCONCEPTIONS THROUGH THE AGES

A REFERENCE GUIDE

Robert E. Krebs

GREENWOOD PRESS
Westport, Connecticut • London

Library of Congress Cataloging-in-Publication Data

Krebs, Robert E., 1922–
    Scientific development and misconceptions through the ages : a
reference guide / Robert E. Krebs.
      p.   cm.
    Includes bibliographical references and index.
    ISBN 0–313–30226–X (alk. paper)
    1. Errors, Scientific—History. 2. Science—History. I. Title.
Q172.5.E77K74   1999
509—dc21   98–5173

British Library Cataloguing in Publication Data is available.

Library of Congress Catalog Card Number: 98–5173
ISBN: 0–313–30226–X

First published in 1999

Greenwood Press, 88 Post Road West, Westport, CT 06881
An imprint of Greenwood Publishing Group, Inc.

Printed in the United States of America

The paper used in this book complies with the
Permanent Paper Standard issued by the National
Information Standards Organization (Z39.48–1984).

10 9 8 7 6 5 4 3 2 1

Contents

Introduction

This book is both a reference work and a presentation of how humans have developed concepts to explain the myriad phenomena we confront everyday.

Throughout history, men and women have come up with many different explanations for what occurs in their environment—both at the macro and micro levels, the seen and the unseen, the known and the unknown. We spend a great deal of time thinking about why things are as they are and then come up with some answers.

Obviously, in the past, not all these answers were correct. Some were beliefs conjured up to satisfy needs that may have been of the real world or of the spirit world. Looking back, many of the answers arrived at were misconceptions about nature and the environment. Based on what we know now, earlier explanations of nature and its forces were irrational. But to early humans, those explanations were rational enough for them to survive and reproduce, and, in the long run, that is what counts in nature.

For thousands of years, people's views of nature were not well organized or thought out. Things were as they seemed, or were believed to be. For thousands of years, most people never considered asking why, how, what, or when. They just accepted whatever they observed and, later, what was told to them or was written by others. The concept of *scientifically* gathering information and arriving at a conclusion was very late in coming.

The scientific method or, more appropriately, the use of the processes of science to answer questions and gain fundamental knowledge is only a few hundred years old. There are still, supposedly, educated people who believe the Earth is flat, that men never landed on the moon, that Elvis lives, that alien spaceships have landed on Earth, and that diseases are not caused by microorganisms.

TYPES OF BELIEFS

Following are several synonyms for *beliefs* that will be used when presenting some of the historic examples of scientific beliefs that have persisted during the ages.

Beliefs can also mean: acceptance and approval; acquiescence and adherence; certainty, conclusion, and confidence; credo, creed, ritual, rule, system; doctrine, dogma; and orthodoxy; faith, fidelity, troth; feeling and fancy; gullibility and naivete; imagination and impression; loyalty and follower; notion, opinion, viewpoint; philosophy and thinking; religion, church, worship, cult; tenet, theory, thought; tradition, trend, trust; and myth, story, folktale.

TYPES OF MISCONCEPTIONS

The term *misconception* in science as used in this book also implies misunderstanding. During the hundreds of years that science was developing, those involved in the enterprise usually did the best they could to understand and describe nature with the knowledge and tools available, and within the political/religious climates of their times. Therefore the terms as used here are not interpreted to mean malevolent or devious deceptions or purposeful mistakes or frauds. Many of the historical scientific misconceptions we will trace have meanings similar to the following synonyms: miscalculation and mismeasurement; miscomprehension and ignorance; misinterpretation; mistake (honest errors); misunderstanding and misconstruing; illusions and delusions; subjective and prejudicial; and subconscious bias.

This does not mean that men and women of modern science have not succumbed to the temptations of deceit or wrong doing to further their careers. Alexander Kohn in his book *False Prophets, Fraud and Error in Science and Medicine* (1988) relates dozens of cases of scientific error, forgery, plagiarism, trimming of controls, cooking, fudging, creating data, and just outright fraud in the basic sciences as well as in biotechnology and medical research. Considering the number of scientists working today the actual number of cases of willful deceit and misconduct is very small yet very disturbing and potentially harmful when used to establish social policy. Mr. Kohn concludes, "Falsification is inevitable in human society, be it scientific or otherwise. Science, like banking, however, has the ways and means to keep misconduct in check" (p. 11).

This book traces the history of scientific developments and how misconceptions arose to explain the universe and nature. Disproving, refuting and, in some cases, building on these early developments and misconceptions is what science always has been about. We start with what we have and add to each bit of knowledge with ever-more refined instruments and methods of inquiry. Science is the only human endeavor that operates with a self-correcting system using controlled experiments. This self-correcting system was only developed over the past few hundred years. In the distant past, using such a system to investigate nature and challenge "truths" was unthinkable. What was known was believed to be the best there was, so one could not easily be a skeptic and question the authority of the "experts" or the God of their time.

This book is organized around the historical developments of the major areas of scientific inquiry. It includes bits of history tracing both the developments and misconceptions of the sciences into the modern era. There are two ways to productively use this reference work: one is to look up your main areas of interest in the table of contents; the other is to seek specific subjects or the names of people in the indices.

Chapter 1

Background: A Short History of Science

We cannot know for sure just when science originated, but when we consider the history of science, we do so with a different perspective than did early humans. We still do not know how men and women, before recorded history, saw themselves in relation to nature and their environment, and what questions, if any, they were asking. During the pre-Christian era people answered such questions with folktales or myths. Sometimes their answers involved reasonably accurate descriptions of nature. Even today, people continue to ask why and search to satisfy their curiosity.

Early humans were not only at the mercy of the forces of nature but were also dependent on the environment for their well-being and survival. Curious humans long ago began the process of trying to understand and control nature. It is reasonable to assume that *Homo sapiens* (Latin for "man, the wise") learned a great deal about nature by using several of science's simplest methods: observation, recognizing relationships between things and events, and most important, trial and error. Early humans needed explanations for events that occurred. They made up stories, folktales, and myths to explain things and events they could not understand. Early humans also made use of a few of the processes of science to aid their survival, better their lives, and provide more leisure time. These processes included using their superior brains to better observe situations and make judgments. They also recognized cause-and-effect relationships between natural phenomena, for example, clouds and lightning accompany thunder, rain, and floods; the amount of daylight corresponds to the change in seasons; they realized a connection existed between the sources of food for the animals they ate and the availability of those animals.

Early humans also had a natural ability to categorize what they found in nature according to similarities or differences of individual things, plants, or animals. From each group humans could consider a specific item and then make generalizations about it based on what they knew about similar objects. For example, early men and women knew that many plants bear fruit and berries. Through a process of trial and error and observing what animals ate, some berries were grouped as poisonous or not good to eat. Other plants, fruits, and berries were grouped as good and were desirable to eat. Once its general grouping was known, a new plant, fruit, or berry could be stereotyped without using trial and error, and thus could be safely eaten. Today people still stereotype all kinds of specific objects, events, and individuals according to the way they categorize particulars for groups as to real or presumed characteristics.

Being able to establish such general relationships is surely one of the processes of thinking that early humans developed and used. We still use this skill in our everyday lives. As humans developed, such scientific processes enabled them to understand and control their environment. In time, the members of the tribe who were the best hunters became responsible for hunting; other members were responsible for gathering vegetables; whereas others were better suited for duties requiring different types of specializations. Along with the domestication of animals and the development of agriculture, specialization provided opportunities for people to do more than just survive. They could now spend more time learning about nature and develop civilizations, with the resulting development of the arts, government, and religion.

Curiosity aside, men and women wished to understand the land, the animals, the heavens, and the relationships of these to themselves. Over hundreds of thousands of years, they asked questions about all aspects of nature. But who supplied the answers? Most likely, they consulted a wise leader in their groups, tribes, clans, or nations.

These wise leaders felt obligated to provide explanations. Because followers were not usually knowledgeable enough to dispute any of these theories, the leaders developed stories to provide some explanations for natural phenomena. These stories became myths and folktales. Some stories were specific and were incorporated into the culture as firm beliefs. Some of these myths and beliefs are so firmly entrenched that they are still with us today. These beliefs then became common knowledge. Everyone knew them, so they must be true. This common knowledge implied common sense. At least, it made sense to thinking people.

The original meaning of the common sense concept—that it is common, and thus many people possess it—is what led to so many nonrational explanations for natural phenomena. Ironically, although common sense led to many incorrect beliefs and misconceptions of nature, it did not always lead to dead ends, but rather to a continuing curiosity and search for better, more rational answers.

For example, common sense told everyone with eyes that the Earth was flat. Thus, all kinds of incorrect observations, speculations, interpretations, and pronouncements about the motion of the planets, the sun, and stars were made to explain the common sense concept of an unmoving, flat Earth.

When someone made observations that provided an interpretation that explained how the Earth moves within the heavens, these rare and unique persons were considered heretics, not people with unique *un*common sense. Over the ages, those few with uncommon sense were the people who made the greatest strides in our understanding of ourselves and the universe. They are the great thinkers and scientists we have all read or heard about, including the three most influential Western philosophers from Greece: Socrates, Plato, and Aristotle. Many historians state that it was the Greeks who invented science as an intellectual process.

What we have been discussing so far is considered "science as a continuous process." Over the ages we have learned a little at a time. Some knowledge was found useful, some not so useful. Not only did knowledge accumulate but so did the methods, techniques and instruments for gaining knowledge. (Which came first, the knowledge or methods and instruments of inquiry?) The knowledge, science, and technology of one age built on and added to the knowledge, science, and technology of past ages. Sometimes incorrect knowledge is discarded. Sometimes it withstands the tests of trial and time. Regardless, knowledge and information accumulate. The processes for gaining knowledge accelerate. One such period of rapid scientific growth occurred during the Renaissance, when people learned how to ask questions for which rational answers could be expected (hypotheses) and arrived at answers through controlled experiments.

Many intellectual disciplines propose theories, accumulate knowledge, and progress in stages over time. Science is the *only* discipline that uses a unique built-in self-correcting mechanism that oversees the accuracy of the results obtained. An incorrect bit of knowledge, theory, hypothesis, experimental result, or an erroneous conclusion cannot long survive the experimental, intellectual, and persistent assault of fellow scientists.

At the same time, it is important to realize that discoveries and beliefs and even some scientific misconceptions of the past *did* lead somewhere. The ancients' ideas were not all dead ends, they were not all mistakes, and they certainly were not all conscious deceptions. Past scientists' misconceptions about nature were the best possible explanations they could make with the limited resources available at the time. As Sir Isaac Newton (1642–1727) said, "If I have seen a little further it is by standing on the shoulders of Giants" (Asimov, 1988, p. 295).

There is another theory as to how science developed over the ages. Alan Cromer (1993) in his book, *Uncommon Sense: The Heretical Nature of Science,* postulates that science was not a natural sequence and accumulation of invention, discovery, and knowledge from ancient to modern times. His position is that science had several starts, pauses or interruptions, followed by new beginnings. He expresses the theory that science as we know it developed in early Greece, possibly because of the intellectual culture and tolerance for objective inquiry and debate that existed there (in contrast to the early Roman culture, which emphasized oratory rather than intellect). The Greeks were the first to formalize the concept of generalization, in which a theory or idea of nature was established and then added to using specific observations that addressed the practical aspects of nature. Science, or rather technology, spread eastward from Greece to Babylonia, Egypt, and also to India and China. Technology, without the exploration or conceptualization of science, flourished in these Eastern countries for many centuries, as Grecian civilization declined and Rome was consecutively destroyed by the Goths, Vandals, Franks, and Norsemen.

The Arab countries (Persia, now Iran, and other areas of the Mideast) became involved in many wars. Muslims conquered India and then spread their conquests westward. In time, this Arab expansion brought the knowledge of science and technology back to the Mediterranean region as they conquered countries along the way. Not all historians agree to the degree that the Muslim world brought science back to Greece, Rome, and Egypt, but there is some evidence that the Muslims carried science as far as Spain, where their conquests ended. From here science spread to many countries throughout Europe and finally to the New World. According to Jones and Wilson (1995) in *An Incomplete Education,* the Arabs' contribution to science was not in developing new scientific concepts, but rather it was the Arabic language that played a role in spreading knowledge of science from India to the Mediterranean and Latin countries and finally to Spain, and hence to Europe. As Jones and Wilson say (p. 545), "Islam made science international."

In other words, knowledge of science and technology moved around and found fertile ground for growth. At times, this growth was thwarted, such as during the Dark and Middle Ages (from about the fifth to the fifteenth centuries). The major spurts of growth were during the Renaissance (1450 to 1650), and from the middle of the seventeenth century to the middle of the nineteenth century. Of course, great advances in science occurred from the nineteenth to the twentieth centuries, particularly over the past 50 years.

There are a number of scientists, mostly theoretical physicists, who believe that this most recent growth spurt over the past 100 years or so may be the last such period of rapid growth of science and technology. They believe that the major scientific laws, principles, and theories are now known and that today scientists are just refining their theories and hypotheses and simplifying their concepts. They postulate that the future source of answers about nature and the universe will rely on metaphysics, philosophy, or theology. Even so, most scientists, as well as most people, do not believe the proposition that science is "dead." There is still much we do not know about ourselves and the universe.

WHAT SCIENCE IS

There are many definitions and descriptions of science. Some depict the methods, processes, or intellectual activities of science, and others, including many advertisements, use the word "science" in a very general way when referring to specific products. In fact, the term "science," when used in the popular sense, usually lacks any connection with the actual intellectual enterprise that science really is.

In his book *What Is Science?*, Norman Campbell (1953) points out one classification:

The two forms, practical and pure science, are probably familiar to everyone; for the necessity for both of them is often pressed on the public attention. . . . Students of pure science denounce those who insist on its practical value as base-minded materialists, blind to all the higher issues of life; in their turn they are denounced as academic and unpractical dreamers, ignorant of all the real needs of the world. If the two forms of science were really inconsistent with each other, both sides could present a strong case. (pp. 1–2)

Campbell continues by saying that both pure and practical scientists are needed. Pure science is usually carried out by those few of us who possess the inclination and necessary intellectual talents, and who enjoy the rigors of abstract thought. These individuals feel that these human qualities make us more than brutes. The practical sciences (also known as applied science or technology) provide the material benefits that enhance the ways we live and work, and they provide the freedom necessary for developing our higher interests, including the arts and theoretical sciences. In no way is practical science inferior to abstract learning—they are inseparable and often form two different approaches to similar problems, and both use many of the same processes or methods of science.

Several definitions of science are found in *Webster's II New Riverside University Dictionary* (1994):

1. a. The observation, identification, description, experimental investigation, and theoretical explanation of natural phenomena;

   b. Such activity restricted to a class of natural phenomena;

   c. Such activity applied to any class of phenomena.

2. Methodological activity, discipline, or study.

3. An activity that appears to require study and method.

4. Knowledge, especially that is gained through experience.

Definitions 2, 3, and 4 are not considered adequate definitions for pure science because they apply to just about any type of human intellectual endeavor. For instance, empirical or experientially gained knowledge does not involve the kernel or core of science, which is the controlled experiment.

The first definition provides several crucial activities (processes) related to science. The first three activities in this definition—observation, identification, and description—are important to any intellectual examination. These three processes (and others) along with the fourth and fifth terms in the first definition, that is, *experimental investigation and theoretical explanation of natural phenomena* are essential to the sciences that involve controlled experimental research. Scientific research implies that a controlled situation will be part of the investigation, which the scientist hopes will lead to new knowledge—what pure science is all about. When a controlled experiment provides results in conflict with the stated hypotheses the entire experiment is not considered a "failure." If a specific and concise hypothesis does not have some chance of being wrong it has little utility. After all, a hypothesis stated in a very general way proves nothing because there are few exceptions to any very broad statement. In addition, if the hypothesis fails, some knowledge is gained as to what does not work, which may lead to new approaches and hypotheses related to the problem.

The experiment is the main distinction between the pure or basic sciences and the applied sciences, and between science and nonscience. The controlled experiment, which is unique to many of the basic sciences, also separates these sciences from other areas of study or research, for example, historical research, biblical research, parapsychology, philosophy, economics, and so forth. This in no way takes away the importance of historical and library research, or empirically gained information, which, when carried out in a systematic manner, can lead to much knowledge and insight. Because we cannot accurately reconstruct the past, we are limited to reading, hearing about, or indirectly learning history from artifacts.

Several historians point out that modern science is based on two very important achievements of Western civilization: The invention of a formal system of logic by the Greeks (e.g., Aristotle and Euclid); and the Renaissance (fifteenth to the seventeenth centuries), when people became more interested in learning more about humans and nature than was provided by the old myths. During this period a great discovery was made. People learned a new way of thinking and arriving at knowledge. In essence, we now think of this as a refinement of the concepts of cause and effect and correlation, that is, the recognition of relationships between natural substances and natural processes (matter and energy) through the use of systematic observations and controlled experimentation. Knowing how to ask the right questions of nature and designing experiments to eliminate incorrect answers so that valid and verifiable scientific theories and laws could be developed was a giant step for science. This was also the era of the development of mathematics, statistics, and probability, which were used in making measurements and predictions. The way nature works, it is impossible to make precise predictions of what will happen in given situations. This unpredictability is the result of the many variables involved in experimentation such as randomness and time/geography, which prevent accurate predictions. To arrive at reasonable answers to complex questions, scientists developed statistical averages and probabilities.

Another way to look at how science developed over the ages is to consider how theories are either confirmed or discarded. True scientists seek out patterns in the natural world, propose concise hypotheses (questions) that have some chance of being wrong, and then test their hypotheses by experimentation. When many scientists repeat the same (or similar) experiments, and a hypothesis is not refuted, in time it may become a theory. Conversely, if the experimental test contradicts the hypothesis, or if other scientists cannot get the same experimental results, the theory is rejected. A failed experimental proof of a hypothesis is not a calamity, it is just part of the scientific process.

Of some interest is the concept of scientific laws, which may be arrived at by theory, philosophy, or observation. A scientific law is a universal statement of relationships between or among natural phenomena that always occur in the same way under the same conditions for the entire universe. Consider the law of gravity: Bodies of matter always fall toward the Earth, not away from it. Or more scientifically, any two massive bodies are attracted to each other in direct proportion to the product of their masses and in indirect proportion to the square of the distance between them. The Scottish philosopher David Hume (1711–1776) stated that when our ideas and theories lead to natural laws it is because nature is lawful. In other words, nature is predictable. It is up to us to figure out how to make our predictions accurate and honest as related to how nature really *is*. Science can be thought of as both a product and a process, or as a noun and a verb. We often think of science as the end product (noun), which can be either our knowledge base or content, and the material benefits derived from that knowledge. But, more important, science is also the methods or processes (verb) we use to arrive at the knowledge that leads to the product. Charles Singer in his book, *A History of Scientific Ideas* (1966, p. 2), refers to science as the process of making knowledge: "By derivation *scientific* implies *knowledge making,* and no body of doctrine which is *growing,* which is not actually *in the making* can long retain the attributes of science."

There is no one scientific method or procedure used by all scientists. Scientists do not always start at step 1 and proceed to the final step in a deliberate sequence. If they have adequate data, knowledge, and background, they may follow some other sequence. Or they may make several starts, find what they are doing is unproductive, back up and try something else. Or they may continue the investigation, arrive at conflicting, erroneous, or misleading data, which can result in the type of scientific misconception we have been discussing. Regardless of where scientists start a research investigation, the controlled experiment and critical observations are the most crucial processes of any scientific method.

Some examples of the processes that are so important to scientific research and investigations are as follows:

> *Observing*—The thread woven throughout scientific investigations;
>
> *Critical thinking*—Logic, theorizing, induction/deduction;
>
> *Curiosity*—Wanting to know, recognizing solvable problems;
>
> *Relationships*—The association of or relationships between old and new knowledge. Association between cause and effect;
>
> *Forming hypotheses*—Asking answerable questions;
>
> *Controlling variables*—Experimentation, research, testing;
>
> *Collecting and treating data*—Organizing and using input;
>
> *Communicating*—Keeping notes, recording progress and results, organizing output, and publishing findings.

There are several other ways people think of science. *Descriptive science* explains how the products of science, as well as technology, are conceived and reported by the media for popular consumption. Most people learn science by being told or reading or hearing about the products of science—not the methods or processes of science.

Theoretical science provides explanations of natural phenomena. Expounding a scientific theory often requires a grounding in higher mathematics, which may hinder most of us from coming up with new scientific theories. Formulating a theory is an intellectual activity that may precede using the processes of science to provide hypotheses that will be experimentally tested to provide evidence that will either back up the theory or send it to the dustbin of history. A theory can also result from the procedure just described, rather than precede it.

It is important for all of us to be able to distinguish between science and pseudoscience, between truth and error, between reality (what is) and irreality (what is not), and between the rational and nonrational.

WHAT SCIENCE IS NOT

Aristotle (384–322 B.C.) made the following statement that epitomized "what science is not": "The worst form of inequality is to try to make unequal things equal." It is unfortunate that people use the words "science," "scientific," "scientific research," and so on in ways that have no meaning related to science as a controlled experimental process. For instance, the "science" of exhibiting an athletic or music ability or skill, the "science" of biblical research, the "science" of the psyche, the "science" of picking lottery numbers, and so forth. It is just as critical to recognize what science is not as it is to understand what science is.

There are several ways to classify sciences and nonsciences. One distinction we have just made is between basic or pure science and applied or practical science. (Please note that not everyone would agree with the following classifications, they are used for illustrative purposes.) Some examples of basic and applied science follow:

Basic Sciences (experimentally based)	Applied Sciences (empirically based)
Biological sciences	Economics
Microbiology, Botany,	Engineering
Zoology, Genetics, etc.	Astronomy
Chemical sciences	Anthropology
Organic, Inorganic,	Mathematics
Physical, Nuclear, etc.	Psychology
Physical sciences	Medicine
Energy, Particle physics,	Sociology
Nuclear physics, etc.	Paleontology
Earth/Space sciences	Archeology
Cosmology, Geology	

Another distinction can be found between science and nonscience, or as it is sometimes called, borderline science, near science, and pseudoscience, which may be related to superstitions and magic. Some examples follow:

Borderline Science	Pseudoscience/Superstition
Psychiatry	Astrology/Horoscopes
Political science	Alchemy
Social science	Parapsychology
Psychology	Extrasensory perception, etc.
Philosophy	Theology/Mythology
Epistemology	Crystals/Small pyramids, etc.
History	

There are several other important concepts to help us understand what science is not. Confusing science with these concepts in the past often led to misconceptions.

Science Is Not *Democratic*

We may vote on what specific science undertakings to support with our tax dollars, or on what philosophical ideologies will lead our nation. We do not vote on what theories, hypotheses, methods, data, and the results of experiments scientists will provide to explain natural phenomena.

We do vote or decide, however, on how these results (products of science) may be used. This is not science, but rather politics and government, economics and commerce.

Science Is Not Based on Everyone's *Opinion(s)*

Opinions, by nature, are arbitrary and subjective. Science and nature are not. Although everyone should be entitled to opinions, not everyone's opinions do or should carry the same weight of authority or be considered equally valid. There are some persons more informed than others, who are also wiser and better trained to explore nature, the universe, social phenomena, economics, history, and so forth. It is expected that these more informed and knowledgeable people can arrive honestly at logical, rational, and sensible solutions for problems related to their areas of expertise. "Every man has a right to his opinion, but no man has a right to be wrong in his facts" (Montapert, 1964, p. 145).

Science Is Less *Subjective* Than Other Academic Specialties

When a person makes a judgment based on his or her personal opinions, beliefs or experiences, the statement is most likely to be *subjective*. If the judgment is not based on personal beliefs, emotions, biases, and prejudices, but rather on measurable and verifiable (repeatable) observations, the statement may then be considered *objective*.

The distinction between being subjective and objective is somewhat related to the value and accuracy of one's opinion. On a scale of 1 to 10:

| 1 | 2 | 3 | 4 | 5 | 6 | 7 | 8 | 9 | 10 |

Subjective
(irrational; biased)

Objective
(rational; unbiased)

Most of us fall somewhere near the middle or lower end of this scale.

It is important to note that even the greatest thinkers and scientists are not perfect "10s." An "8" or "9" is a goal for which we should all strive, but may not reach as all of us are human beings (including scientists) with personalities, opinions, prejudices, and biases. By nature we are subjective beings, that is, we each have our own individual minds and personalities.

How does one measure something that is abstract? It will have no height, length, width, weight, color, and so on. For example, the concepts of success or beauty are abstract ideas and are thus based on individuals' subjective opinions, and may be defined, but not measured without bias. In comparison, it is not difficult to measure a brick. It has length, width, depth, weight, density, color, temperature, and so on, which makes the concept of a brick less abstract and more concrete, and thus one's concept of a brick is more objective and may be measured with less bias. It is unlikely that you will get many people to agree on what success or beauty are, or how to measure these abstractions, but once they all see a brick and define it, you won't get much disagreement on its measurable characteristics or what it is.

Science Is Not *Magic*

There are at least two definitions for magic. First, it is sometimes called the art of forecasting natural events or forces using the supernatural. Or it may be thought of as attempting to produce supernatural effects to control natural events. Second, magic is the exercise of sleight of hand or trying to mystify or enchant for entertainment.

Obviously, neither definition begins to use any of the processes of science, and in particular, magic will not and cannot tolerate any type of controlled experiment. Related to magic as well as to the supernatural is the occult, which is derived from the Latin word *occultus,* meaning secret. Also related to magic are sorcery, voodoo, witchcraft, wizardry, and necromancy; none of which use the methods of science nor have any scientific basis.

Science Is Not *Myth* or *Superstition*

Let's start with definitions for both of these nonsciences. Myths or mythology developed in prehistorical or preliterate times as storytelling designed to explain the unknown, both the natural and the supernatural. Many myths became part of a groups' culture and expressed commonly felt emotions. In essence, myths are fictitious stories or folktales, which, no doubt, served a purpose for people when they could not make sense of their humanness or their environment. Myths, no doubt, aided humans in coping with the unknown.

Superstitions are beliefs that form behavior and effect one's actions. Myths served a primordial purpose, whereas superstitions are based on magic (deception), the occult (secrecy), and ignorance (unknown). Superstitions are practices, rites, or beliefs that exist despite volumes of evidence that clearly indicate that they are invalid. Superstitions are a form of irrationality that ignore the laws of nature. Superstitions evolved into beliefs to address peoples fears, particularly fear of the unknown.

Science Is Not the *Paranormal*

The paranormal, also described as pseudoscience, is often confused with science. Paranormal activity, which is sometimes studied in great depth, is not science because it does not use a system of controlled experimentation that thrives on error. Many of the paranormal activities are based on ancient beliefs, superstitions, and myths that have been brought into the modern world as nonscientific misconceptions.

The distinction between paranormal phenomena and science is that science is based on observed facts, repeated and verified observations, as well as "tests" that determine the validity of observed events, while the paranormal cannot hold up to this type of scrutiny and close examination. Real sciences, or exact sciences, eliminate errors as they are recognized. Although many believe in the validity of the paranormal, these phenomena cannot be considered exact sciences. Also included among these pseudosciences would be frauds, hoaxes, and curses. Examples of the paranormal are:

1. *Parapsychology* (not explainable by known natural laws)
   - telepathy (nonsensory communication by mystical powers)
   - clairvoyance (power to see beyond one's natural senses)
   - precognition (a form of clairvoyance that knows events before they occur)
   - dreams of future events (a form of precognition)
   - psychokinesis (moving remote objects by psychic powers)

2. *Extrasensory Perception—ESP* (supernatural perception)
   • visions, ghosts, aberrations, calling up spirits, psychic predictions

3. *Necromancy* (communicating with the dead to predict the future)
   • channeling (past souls reentering present humans)
   • reincarnation (being reborn from a past life or spirit)
   • resurrection (rising from the dead, from mythologies)

Science Is Not *Religion*

This example of what science *is not* is the most difficult to distinguish because of our own beliefs and how we accept our faith's implications to arrive at explanations for humans, life, and nature not everyone agrees upon; but religion and science are not necessarily opposed to each other. They are just different human enterprises that try to answer many of the questions about ourselves and our universe. Science provides us with rational and accurate answers to questions that can be tested, as well as predictions about the "hows" of nature. Religion cannot be tested in a similar fashion, and does not have a built-in error-correcting system, but it does provide answers to the "whys" of nature that satisfy many people. This "how" and "why" concept is one of the main difference between science and theology.

In addition, any and all questions are okay to ask in science if there is some possibility of arriving at a quantitative answer. Skepticism is accepted, even encouraged. Inquiring, skeptical minds are not always tolerated in other areas of human intellectual activity. Science requires some form of evidence for one's particular predictions. But the same types of proofs required for science are not a requirement for religious history or prophecies. Some degree of faith is required for all types of human activity that searches for answers, however, regardless of the methods or beliefs that are used to seek out those answers.

Scientists, in one sense, have a great deal of faith, or rather, they believe in the methods they use to arrive at knowledge, and once verified and reproducible, that this knowledge is accepted as it relates to the real world. That is, at least until something different that is even more verifiable and reproducible comes along and provides an even more rational explanation of the phenomenon.

Religion is based on faith, which is the belief in something for which there is no physical evidence. Religions require a degree of acceptance through faith in their theologies, histories, prophecies, tenets, beliefs, and so on. There is some, but not much, room for change and growth in theology and religious beliefs over long periods of history, even though these beliefs are unverifiable.

Another interesting parallel between science and religion is the concept of "revelation." Institutionalized religions of all varieties base much of their theology and written tenets (holy books) on what is sometimes referred to as *divine revelation* or the *word of God*. A similar concept exists in science, though it usually is not attributed to a supernatural entity. There are many examples of knowledgeable and well-prepared scientists who, through insight or revelation, came up with a very uncommon sense answer for a problem they were working on. Most scientists are professionally and intellectually prepared to make use of a revelation or inspiration when it occurs.

Science Is Not *Politics*

In the introduction to his book *The Tempting of America: The Political Seduction of the Law,* Robert H. Bork (1989, p. 1) states: "[P]olitics invariably tries to dominate another discipline, to capture and use it for politics' own purposes, while the second subject—law, religion, literature, economics, science, journalism, or whatever—struggles to maintain its independence."

Carl Sagan places equal responsibilities on scientists and politicians as to how science should be used for the welfare of humankind. In his book *The Demon-Haunted World: Science as a Candle in the Dark* (1995, pp. 11–12), he says: "The sword of science is double-edged. Its awesome power forces on all of us, including politicians, but of course especially on scientists, a new responsibility—more attention to the long-term consequences of technology, a global and transgenerational perspective, and incentive to avoid easy appeals to nationalism and chauvinism. Mistakes are becoming too expensive."

Several differences can be seen between politics (government) and science:

1. Politics deals with abstracts differently than does science, that is, subjectively versus objectively.

2. Politics does not need to "prove" anything, whereas science is based on proof.

3. Politics does not need to consider validity/reliability for its policies, whereas science insists on being able to get accurate and repeatable data and results.

4. Politics is concerned with "predictions" and policies that are often based on insufficient and inaccurate data, whereas science makes few predictions without adequate and accurate data.

5. Politics is based on making others "happy" and politicians' self-preservation, whereas science is based on questioning preserved concepts. Scientists are also concerned with self-preservation, but their concern is usually centered on their scientific discipline so it can better serve society.

6. Politics is based on control of human activities and economic/social endeavors, whereas science is concerned with understanding our activities and the makeup of nature.

One conflict, not generally realized, is that scientists have a healthy mistrust for pronouncements and arguments from authorities, particularly self-professed authorities, including some of their colleagues. This includes political authorities at any level of government who are not prone to prove, or be responsible for the outcomes and results of their contentions, predictions, or policies. It also includes scientific authorities who do not have adequate, accurate, reproducible data before they publish the results of an experiment in refereed scientific journals. Although this could lead to a degree of superiority in scientists, it does not absolve those in political power from the responsibility of formulating policies on the basis of the best science available, or of substantiating their policies, and justifying the consequences resulting from those policies, which can seriously affect all of us.

Democracy is one form of government where such conflict is minimal. Science and democracy have many concomitant values. They both believe in freedom of information and ideas. They both reject secrecy. They both encourage debate, use of reason, and will tolerate and even encourage unusual ideas. Both democracy and science have "baloney detecting kits," which Carl Sagan (1995, pp. 209–11) considers special methods for eliminating nonsense. Even so, it is the nature of all government to desire influence and control over the direction and independence of science.

Let us consider several examples of this attempt at control. The Italian philosopher Giordano Bruno (1548–1600) proclaimed that there were many many stars in the sky, that the sky was infinite, and that the Earth moved in the heavens. This did not coincide with the concepts and dogma of the existing political powers of his day. Because Bruno would not recant his ideas he was burned at the stake—in the year 1600! If this political climate still existed it could put quite a chill on scientific thought, research, and writings.

Another similar example involves Galileo Galilei (1564–1642), who accepted the Copernican theory that the center of our solar system is the sun, not the Earth. This did not set well with the powers of the Church, who accepted Ptolemy's theory that the Earth was the center of the universe. Galileo was subjected to the Inquisition and threatened with torture and death. Being 70 years old, he recanted and spent the rest of his life in prison.

The following chapters will trace the historical aspects of scientific developments about nature and the ideas that guided men and women in their search for understanding of how things came to be and how things worked. As we will see, many scientific misunderstandings and misconceptions were part of the process of scientific progress. Many of these misconceptions resulted from people's lack of knowledge, a lack of methods for gaining knowledge, and the lack of tools and instruments to better measure and verify observations, justify predictions, and answer questions. Over the centuries, curious people pondered and grappled with what it was "all about."

Chapter 2

Medicine and Health

INTRODUCTION

Even before the era of recorded history, men and women were surely aware of and interested in their bodies, illnesses, and how to alleviate pain and suffering. They did not have the benefit of modern medicine, drugs, or health care systems. As a consequence, their lives were short and mean.

Even though we do not have a formal written record of early people's medical problems, we do have fossils and other types of physical records. For instance, evidence of arthritis has been found in the joints of bones of ancient men and women (as well as prehistoric animals). There is evidence that one type of cure called *trephining* was attempted by drilling holes in skulls of patients either to let evil spirits causing insanity escape, or to relieve headaches or other illnesses. Evidence of medical practices are found on cave pictographs, which show a variety of medical procedures. Additionally, surgical tools have been located among the remains of prehistoric men and women.

There were two basic approaches to early medical practices. One was based on the development of superstitions based on early man's efforts to explain nature and his own existence, to avoid evils, and regulate his fate. To cure sicknesses early people attacked the cause of the disease as they saw it—the evil spirits. These types of diseases were treated by trying to get rid of the malevolent demon by magic, dancing, chants, and talismans (magic charms), or by witch doctors performing incantations. Because the patient was obviously invaded by evil spirits, it was necessary to rid the patient of this malady. In addition to all the magic, dancing, and so on that was external to the patient, he or she was often beaten, starved, forced to vomit or bled to extract the spirits.

The second approach was more successful. Early people learned how to use surgery and other techniques to help the body self-heal wounds and broken bones. These techniques did not require the extraction of evil spirits, but still involved many superstitions based on erroneous perceptions. There is some evidence that early men and women knew how to clean and suture wounds, apply poultices, and seal off bleeding with pressure and burning (cautery). They also had some experience in setting broken bones and joint dislocation.

In addition, early men and women turned to the plant world to cure their ills. Medicine women and witch doctors learned from experience which plants could be used for therapy as laxatives, emetics, diuretics, and anesthetics. We still use dozens of these same plants for medicinal purposes. One such example is the foxglove, from which we extract digitalis, a drug used to treat heart disease.

Imhotep (c.2725 B.C.) of Egypt is the oldest physician/healer on record. He is generally accepted as the founder of medicine. Imhotep was the physician and counselor to King Zoser, also known as Djoser (c.2686 B.C.). In ancient Egypt medicine was changing from the magic and religious approach to more observational diagnoses and treatments based on experience. Egyptian physicians spent many years studying and training. They predate our medical education system by many hundreds of years.

In one form or another the practice of medicine existed in many geographic areas and cultures over the Earth for many, many centuries. It is amazing how similar were the beliefs and misconceptions throughout these ancient cultures. Today, as in times past, the essence of the practice of medicine is a matter of observation, opinion, and experience. We are fortunate to have the historical experiences of past scientists and physicians to call on so as to expand our current medical practices.

It was not until about the seventh or sixth century B.C. that the magic-spirit concept gave way to medicine based on astrology and alchemy. This advance involved some observations and experience, but still was related to unknown powers.

The earliest records we have of peoples' concern for medicine comes from mythology. Hermes, the winged-footed Greek god, traded his lyre and shepherd's pipe to the sun god Apollo for the golden shepherd's staff with two snakes entwined around it, which was called the caduceus (see Figure 2.1). Mercury, the Roman coun-

Figure 2.1 Mercury's caduceus with two entwined snakes, which has become the symbol of medicine.

terpart of Hermes, is often confused with the healer Asclepius, the god of medicine and patron of Greek physicians. Asclepius, who was Apollo's son, cured people by appearing in their dreams. Asclepius was raised by the cen-taur Chiron, who taught him all about the art of healing. Asclepius also had a caduceus, but with just one snake wrapped around the staff (see Figure 2.2). As the myth goes, Asclepius raised a patient from the dead. This angered Zeus, who killed him with a lightning bolt. For centuries the sick and lame visited the temples built to honor Asclepius hoping to be cured, just as they still do today at some Christian shrines.

Figure 2.2 Asclepius's caduceus.

Medicine developed slowly in several different countries. It is not clear from the archaeological findings just where it started—Asia, the Middle East, or even southern France. There is evidence of early practices of medicine in the Asian civilizations (India and China), the Mesopotamian countries (Sumerian civilization, Assyria and Babylonia), the Middle East (ancient Israel and Persia or Iran), Egypt, Greece/Rome, and southeastern Europe.

Scientific progress was slow, covering about 2000 years, until the seventeenth century when the concept of the scientific method of conducting controlled experiments combined with objective observations drove out alchemy, magic, and myths. From the 1700s the development of medicine was slow but steady, until the twentieth century. Only since the 1950s have the sciences of biology, genetics, chemistry, physics, and the development of medical instrumentation rapidly accelerated our knowledge of medicine. Even so, from the earliest times to the present, the practice of medicine is and has been considered an art. It is the art of diagnosing, treating, and preventing disease based on keen observations, in-depth knowledge, the application of science and medical technology, and practical experience. Although the practice of medicine is not itself a science, it is now a highly skilled profession based on the best scientific knowledge and technological instrumentation known to the world.

This chapter traces some of the early, as well as some modern, scientific developments, beliefs, and misconceptions related to medicine, anatomy, physiology, disease, and health. Some of these beliefs and misconceptions may overlap as a result of the timing and nature of the developments in history.

ANATOMY AND PHYSIOLOGY

The word "anatomy" is derived from the Greek word *anatome,* which means dissection or to cut. Anatomy can be thought of as the structure of something—how it is built and how it is put together, whereas physiology is how that thing works, for example, how the organs function.

An analogy is found in the field of architecture, which will help to clarify the reasoning of examining the two together. The architect Louis Henri Sullivan (1856–1924) stated that buildings built for people should be an expression of the architect as well as designed for a purpose, that is, "the organism in its surroundings" (Twombly, 1986, p. 226). Sullivan also expressed the concept that architecture should appeal to life's or nature's three main phases, namely, growth, decay, and death (p. 225). His most notable statement grew out of the following concept: Architecture is preeminently the art of significant forms in space—that is, forms significant of their functions (Bragdon, 1931). This led to Sullivan's famous dictum that has been interpreted as: *Form ever follows function.* Carl Sagan and Ann Druyan state in their book, *Shadows of Forgotten Ancestors* (1992), that at the micro level or molecular level this macro concept of *form follows function* is incorrect. An example of *their* concept of function follows form at the molecular level is the A, C, G, and T nucleotides of the DNA molecules. In other words, the functions preformed by the nucleotides of DNA are dependent on the form of their arrangement in the DNA sequence.

For plants and animals to function as total living organisms, it is necessary that their forms (anatomy) be designed to accommodate their organs' specific functions (physiology). For instance, the circulatory system could not perform its function in mammals if it were not for its particular structure and its relationship to other anatomical structures, such as, the skeleton, brain, and lungs. In other words, living organisms follow the dictum of "form follows function," which architects, artists, and others have recognized as a synergism of nature and as such applied the concept to their crafts. The study of physiology and anatomy is the basis for all medicine.

History of Anatomy and Physiology

Most likely, early people learned the basic structure of the body and some of the functions of different organs by experience. While dressing the carcasses of animals killed for food, they could not but recognize the hearts and relate this organ to the flow of blood, or observe how the skeleton provided the necessary support for the animal's weight. Prehistoric men and women may have also tended the wounds and bone fractures of warriors and, no doubt, could see the life ebb from sacrificed animals and humans as the hearts were removed. The only records we have of the earliest knowledge of anatomy and physiology are cave pictographs and some instruments that seem to have been used for surgery. In addition, a number of fossilized skeletons have been found that indicate bone fractures were repaired.

As mentioned previously, the earliest physician known by name was the Egyptian astrologer/magician/physician Imhotep (c.2725 B.C.). Egyptian physicians conducted examinations of their patients, prescribed drugs, and tried different cures. The Egyptians did little to advance knowledge of anatomy, except for what they learned through the practice of embalming. They also did some minor surgery and practiced dentistry.

The ancient Egyptians performed circumcisions, more as a means of identifying slaves than for health or religious reasons. They performed total castrations to produce eunuchs who protected the harems. When limbs became infected or gangrenous, amputations were performed. If someone injured an eye, eye operations were tried. Egyptian physicians also performed surgery to correct bone deformities and to remove bladder stones. They even performed some plastic surgery for cosmetic as well as medical purposes. Most of the medicine of this period was of an emergency nature—they did what they thought was required to save lives—even if the procedures in the end killed the patients. For instance, they had no concept of why surgical procedures produced infections that often killed patients. They had no way of anesthetizing patients to relieve pain and suffering, even while amputating a leg. Their knowledge of how the body was structured and functioned was very limited, more than likely they related its processes to mysticism.

Aristotle (384–322 B.C.) was the son of a Greek physician, who at the age of 17 went to Athens to study under Plato. He was never a practicing physician, but became a great philosopher and scientist who covered the entire range of knowledge. Aristotle studied the human body as part of his system of philosophical inquiry, as did others of that time, including Plato and Xenophon. By dissecting animals he made many contributions to the study of anatomy and is considered one of the fathers of comparative anatomy. One of his main contributions was the way in which he would observe things and then classify them according to specific characteristics. His other major contributions were in the areas of logic, psychology, ethics, natural science, botany, zoology, astronomy, meteorology, and physics.

A major concept of Aristotelian thinking was that any circular motion could only be caused by celestial matter and celestial forces. One of Aristotle's misconceptions was that on Earth motion occurred in a straight line, that is, it had to have starting and ending points. When whatever was pushing the object in a straight line stopped pushing (force), the object no longer moved. The idea that circular motion did not exist on Earth made the recognition and acceptance of blood circulating in the body unthinkable. This mistaken concept influenced philosophers, scientists, and physicians for many centuries.

Aristotle formulated a new way of thinking about things, which still influences us today. Later, his approach would lead to the scientific method of asking and answering questions about nature. Aristotle was the teacher of Alexander, who established great libraries and universities in the city named for him. This center of learning influenced not just medicine, but the development of science for several centuries.

Another misconception that persisted over many centuries was that things having to do with the human body, mind, spirit, and the gods always came in threes. Aristotle believed that living things were divided into three types: vegetable—with a vegetable soul, animal—which moved with an animal soul, and man—who was intelligent and possessed a rational soul. The Greek physician Claudius Galen (c.129–c.199, see the section later in this chapter) believed in only three vital functions in man that correspond to the digestive, respiratory, and nervous systems. In the Middle Ages this belief in triads or threes resulted in the limited classifications of living things. For example, animals were organized in their own triad hierarchy of birds, fish, and mammals. Humans had a triad of rank (status), in which angels were above humans, and the Christian Trinity was above all.

The Dark and Middle Ages of Science

The theologian and Spanish physician Michael Servetus (1511–1553, also known as Miguel Serveto), did not accept this triad concept. He mistakenly believed that the human soul was comprised of human blood. This belief later caused him problems because it implied that the soul died with the body and was not eternal. He was, however, the first person to correctly surmise how the heart and lungs worked to purify the blood. Up until this time, most physician/scientists mistakenly believed that the main purpose of the lungs was to cool off the blood. Servetus did not accept many of the triad concepts, including belief in the Christian Trinity. His life was cut short and his writings destroyed when in 1553 he was burned at the stake by Calvinists for his beliefs. His death was meant to be a warning to others not to defy church doctrine.

In the following centuries many countries developed surprisingly similar forms of medicine. Asia (China and India), the Middle East (Babylonia, Egypt, Israel, and Palestine), and the Latin countries (Greece, Italy, and Spain), all made advances and contributed to medical knowledge. Even so, accurate knowledge of the structure and functions of the human body mostly were unknown. From earliest times anatomy consisted of observing dissected plants and animals. Because of religious convictions, many cultures objected to the dissecting of humans, either alive or dead. Therefore, anatomical knowledge was limited to what could be learned through emergency operations, accidents, war, and so on, but very little systematic study of anatomy and physiology was taking place. It took time before scientists recognized the intricate relationships between physiology (functions) and anatomy (form).

Claudius Galen

One of the outstanding early physicians was Claudius Galen, whose writings about medicine influenced others for 1,500 years. Galen had some mistaken conceptions of the anatomy and physiology of the human body. One of his major physiological misconceptions was his belief that if some function of the body changed, an anatomical change of some sort would occur. Conversely, if the body was injured, there would follow a change in some bodily function that was not necessarily related to the injury. Even though his concepts of anatomy and physiology were not always correct he seemed to understand a connection. He was the first to realize that every body organ and its tissues, etc., had been formed for a specific purpose. Galen had an almost spiritual concept of the physiological functions of the human body. He believed the body had three basic functions: (1) the vegetative function, seated in the liver, which provided the blood necessary for nourishment and growth; (2) the animal function, seated in the heart, which acted through the blood and the blood's spirits; and (3) the nervous function, found in the brain, which controlled sensitivity through the fluid in the nerves and through our animal spirits. Galen was an observer of the organs of the body and he tried to figure out how they worked. His misconceptions came from using animals rather than humans as models, the limitations of his observations because of poor instrumentation, and the fact that he tried to relate everything to some mystical power or being. Even today we use animals as surrogates for humans when testing drugs and experimenting with new medical/surgical procedures. We still ignore the fact that animals just do not react as do humans in many of these tests, particularly those for toxic substances (see Chapter 6).

Andreas Vesalius

Many years after Galen, Andreas Vesalius (1514–1564), the son of a famous Belgian pharmacist, became a surgeon and anatomist. He learned much about anatomy by studying the bodies of condemned criminals. He was responsible for the beginning of the end of the dogma of Galenism. Vesalius's dissections and descriptions of the structure of the human body finally caught up with Galen's misconceptions about anatomy. Vesalius was unusual in the sense that he broke with tradition and trusted his own eyes to record what he saw during dissections of human cadavers—not what the Greeks said he should see. Vesalius was the first physician to lay the foundation for modern anatomy. He wrote two famous books, *Six Anatomical Tables* and *Concerning the Structure of the Human Body,* which had accurate engravings of human anatomy. These were the most authentic and comprehensive textbooks of the time, and they corrected over 200 errors made by Galen!

Vesalius's works indicated that he also had many misconceptions about the human body. Because he dissected corpses, not live people, he misjudged the cartilage in the nose for muscle. He mistakenly placed the lens of the eye in the center of the eyeball, not at the front. From his dissection he incorrectly stated that there were only seven cranial nerves, not twelve. He also believed that blood was oxygenated from the liver, not the vena cava (the two main blood vessels between the lungs and heart), and he incorrectly described several bones and joints, including the cartilage in the knee joint. Vesalius underestimated the number of muscles in the body, as well as the different types of muscle; he was unaware of how muscles were affected by nerves. He thought that all muscle tissue was the same. (Depending on what and how you count, there are between 400 and 600 muscles of three main types in humans: smooth, skeletal, and cardiac.) Regardless, Vesalius was far ahead of his contemporaries and made valuable contributions to the field of medicine.

Surgery

Regarding surgery, many misconceptions existed. In ancient times many forms of very general surgery were practiced by various cultures. Because of the lack of knowledge of the human body, misunderstanding of anesthetics and antiseptics, as well as the unavailability of efficient instruments, surgery was a traumatic ordeal for the patient. As today, 2,000 years ago surgery was classified as to the region of the body being operated on. Using techniques such as trephining and removing stones from the scalp to cure insanity, brain surgery was the most advanced type of surgery available at the time. We now know that neither technique cures mental illness. Eye and ear surgeries were also performed when infections developed. Both were crude operations, often leaving the patient blind or deaf. Abdominal surgery, heart, and circulatory system surgery were all dangerous ordeals; the physicians directed the "surgeons," who were usually butchers or barbers by trade. Surgeons did not bother to wash their hands or use clean instruments. Septic conditions and techniques were common. It would take hundreds of years before modern antiseptic surgical procedures were used.

Surgery, without benefit of antiseptics or anesthetics, was practiced on the battlefields of ancient Egypt and Greece to set bone fractures and sew up wounds. Some complicated operations were performed successfully, however. Early Jewish surgeons delivered fetuses via Caesarean section, they repaired intestinal malformations such as anal fistulas and hernias, performed circumcisions, and corrected joint dislocations and bone fractures. Complicated surgery was also performed in India, Egypt, Greece, and Persia. The misconceptions about surgery held in these early times were related somewhat to the early beliefs of the anatomy and physiology of the human body. For example, if patients had a blood disease such as syphilis or anemia they were bled; if they had a mental disease such as delusions or hallucinations holes were cut in their skulls. Surgical practice could not really improve until anatomy and physiology were better understood.

It was not until about the fourth century B.C., with the rise of the great libraries and universities of Alexandria, that systematic anatomical and physiological studies were made. It was at these universities that surgery was first based on more precise study of human anatomy.

The study of human anatomy as a branch of medicine is attributed to the Greek surgeon Herophilus (c.335–280 B.C.). A medical teacher at the University of Alexandria, Herophilus was the first person on record to perform anatomical dissections in public. He correctly noted that blood coursed through the arteries from the heart with a pulsating motion, unlike the blood that returned to the heart through the veins. At about the same time, another physician named Erasistratus (c.300–c.260 B.C.) was able to trace the course of the veins and arteries to the smallest size visible with the naked eye. Erasistratus's major misconception was that air was pulled into the body by the blood as the blood moved down the arteries; and conversely, that air was expelled as the blood moved upward in the veins. Erasistratus also believed that along with the air, the arteries carried "vital spirits" that were distributed to parts of the body. Some of these vital spirits were sent to the brain as animal spirits, and were then distributed to the rest of the body by the nerves.

Erasistratus, who also founded a school of medicine in Kos (Cos), and Herophilus are credited as the first scientists to dissect human bodies to study anatomy, and thus are considered the fathers of anatomy. They came close to describing the circulation of blood in the human body, but were still influenced by the Aristotelian concept that on Earth things moved in straight lines; circular motion only existed in the heavens. Neither grasped the concept of blood circulating in the body. It is assumed that they dissected living humans who were convicted of crimes and were therefore destined to die. At the time, it was illegal to exhume bodies from graves. This taboo against dissecting bodies of the deceased for medical study existed in most cultures until the middle 1800s.

Knowledge of anatomy continued to develop as scientists learned that organs are composed of specialized tissues, which in turn are composed of specialized cells. They began to see how these work together as systems to perform the functions necessary to support the living organism. Between the sixteenth and eighteenth centuries other advancements occurred that allowed for significant improvement in surgical practices. These developments included the invention of the microscope; better surgical instruments; improved anesthesia and antiseptics; and advancements in the understanding of cells, bacteria, infection, and disease. With the development of the cell theory and the germ theory of disease scientists learned how to control many of our health problems (see the section on diseases later in this chapter).

THE CIRCULATORY SYSTEM

History of the Circulatory System

There was little progress in the advancement of knowledge of the structure and function of the circulatory system until the second century after the birth of Christ. Some knowledge did exist before this time, but most beliefs were based on myths and mysticism, not on accurate observations. Therefore, misconceptions about the circulatory system, human anatomy, and physiology were widespread.

The study of the human body did not become a branch of medicine until the time of Hippocrates (c.460–377 B.C.), a Greek physician who is regarded as the father of medicine. Hippocrates both practiced and taught medicine on the Greek island of Kos (also spelled Cos). He is best known for several works that include ethical codes for physicians, what we know of as the Hippocratic Oath. This oath has been revised many times into its present form, which is accepted by the American Medical Association. Hippocrates made many contributions to patient care, as well as advancements in the understanding of anatomy and physiology. However, he and many others held the common misconception that the human body (as well as all matter) was formed of four elements: earth, water, fire, and air (æther). He also believed that these four elements were controlled by hot and cold, wet and dry, and that if the fire element (via calories or heat) was not present, death would follow. Hippocrates died around the time that Aristotle was Plato's student.

There was another Hippocrates (c.430 B.C.) who lived in Chios (another island in the Aegean sea). He was a mathematician who made many contributions to geometry. He compiled an *Elements of Geometry*, which provided the basis for the publication by the better-known mathematician, Euclid.

Aristotle formulated a new way of thinking about things, which still influences us today. Later his approach would lead to the scientific method of asking and answering questions about nature. Aristotle was the teacher of Alexander, who established great libraries and universities in the city named after him. This center of learning influenced not just medicine, but the development of science for several centuries.

Claudius Galen

After Hippocrates, one of the most outstanding early physicians was Claudius Galen, who was born in Greece sometime around the year 129 A.D. Galen dissected many animals to demonstrate how muscles and nerves worked together and how the brain controlled activities, including speech. He described the functions of the kidneys and bladder. His most important contribution was to disprove a belief held for over 400 years that arteries carry air. He demonstrated that arteries carry blood, not air. He also described the differences between the veins and arteries and how the heart and its valves were structured. Galen disproved the earlier concept of the arteries carrying air by tying off the severed artery of a Barbary Ape to stop the bleeding, and then untying the artery to show that the blood again spurted out of the cut artery. Yet, he did not yet grasp the concept of the blood circulating.

Galen held many beliefs about the body that were incorrect. One of his greatest misconceptions was that he considered the liver to be the main organ of the entire cardiovascular system, which is composed of the heart and all the blood vessels in the body. They believed that the liver both made and stored the blood and then sent it to the heart and, from there, to both arteries and veins, to be used up by the flesh of the body. He did not understand that the blood flowed in one direction in the arteries and another in the veins. Nor did he understand the importance of a beating heart. By dissecting the living hearts of animals, Galen was able to identify the correct action of the valves, which control the flow of blood from one section of the heart to another. He identified the left ventricle, which pulsates and expels the blood to the aorta and thus to the body, and the right ventricle, which pumps the blood to the lungs. Galen thought that some blood must leak between these valves so as to distribute the bright red arterial blood and "vital spirits" to all parts of the body. When the fresh blood reached the brain, it could then take the form of "animals spirits" as it passed through the hollow tubes called nerves.

Although he recognized the valves in the heart, his belief concerning the heart's structure was wrong. He thought it was a single pump with two chambers that had small, invisible holes in the wall separating the chambers (septum), which allowed blood to pass from one chamber to the other, and then pump the blood out to both the arteries and veins. Another misconception was that the heart was the source of respiration, that is, when the heart expanded it filled up with air, and when it contracted it sent the air, with blood, through the blood vessels.

Galen also believed erroneously in the concept of "pneuma," which he referred to as three different forms of action in the body. One such action involved sensation and movement and was thought to be centered in the brain; he called this *pneuma psychicon* or animal spirit. The second was centered in the heart and controlled blood and temperature, which he called *pneuma zoticon*. The third physiological action was found in the liver and controlled nutrition and metabolism; he called this *pneuma physicon*. As previously mentioned, another serious misconception held by Galen was that blood flowed from the liver, through veins, to the heart, where it was purified. At the same time, air and vital spirits were added to the blood, which was then pumped to the body and brain. Still another misconception was his belief that the nerves were empty tubes or ducts.

We should, however, give credit where credit is due. Galen was a keen observer who was experienced in recognizing physical symptoms in a patient. He was regarded as an excellent diagnostician and was consulted by many of his colleagues. His errors and misconceptions should not overshadow his contributions to medicine.

Galen's principles, mystical concepts, and writings dominated medicine for hundreds of years. In particular, as related to humans, his concepts of animal anatomy and physiology persisted because of taboos on dissecting humans. As a result of Galen's extensive writings his concepts were accepted as the final word. This acceptance hindered progress in the advancement of anatomy and physiology for almost 1,500 years, well into the Renaissance period. It was not until the sixteenth century that most of Galen's concepts were challenged and his mistaken ideas corrected. This accounts for the statement: "The Tyranny of Galen."

The Dark and Middle Ages of Science

After Galen, medical knowledge declined in Greece, Rome, and Western Europe during the Dark and early Middle Ages. Medical knowledge was kept alive and other contributions made primarily in the East. Chinese, Indian, Arab, and Jewish physicians continued to practice and teach medicine, even as it declined in Europe. With the decline of Rome and Europe came great disasters caused by a variety of diseases. Medical knowledge did not advance much during this time.

Ibn an-Nafis

In about 1240 A.D., an Arab scholar, Ibn an-Nafis, suggested that the heart had two separate chambers, the right and left ventricles, which were separated by a wall of tissue. He got it right. He described (a) how the blood was pumped out of the right ventricle into the outgoing artery, which led to the lungs; (b) how the blood vessels dispersed the blood in the lungs through smaller vessels and how it picked up air; (c) how these smaller vessels formed a large artery that brought the oxygenated blood back to the heart's left ventricle; and (d) after which it is pumped through arteries throughout the body. He still did not know how the blood got from the small blood vessels at the ends of the arteries into the veins to return to the heart.

Renaissance Science

As mentioned, one of the scientists to correct some of Galen's misconceptions was the Spanish theologian/physician, Miguel Servetus (1511–1553), also known as Miguel Serveto. As reported in *The History of Medicine* (Margotta, 1996, p. 100), he published *The Restoration of Christianity* in which he correctly described how the blood circulated from the right ventricle to the lungs, back to the left ventricle, and then was pumped through the arteries. He was one of the first to explore the circulatory system in living humans. He described how the blood was distributed into the tissue by smaller vessels. For his views on religion and the circulation of blood in humans he was burned at the stake by the Calvinists.

William Harvey

In the early 1600s Girolamo Fabricius (1533–1619), an Italian physician, dissected veins in legs and observed small valves that prevented blood from flowing downward or backward in the veins, while allowing blood to flow forward or upward toward the heart as one moved the leg muscles. Fabricius did not publish his work until later because of the great influence Galen's concepts still had over the medical community. An Italian anatomist, Realdo Colombo (1516–1559), was given credit for accurately describing the circulation of blood in the smaller vessels. However, it was the English physician William Harvey (1578–1657) and the Italian physiologist Marcello Malpighi (1628–1694) who really put to rest Galen's misconceptions about the circulatory system. It was Harvey who first proposed that the blood, circulating from the heart through the arteries to small blood vessels, then returned to the heart through the veins.

Instrumentation

Still, one of the great mysteries was how the blood flowing out of the heart into arteries, which ended in very small vessels, was able to get into other small vessels for a return to the heart through the veins. What was the connection? Harvey made a jump in his conclusion that later proved correct. He maintained that there must be other very small vessels, too small to see, that were the bridges between the ends of arterial and venous blood flow.

This problem was solved by Marcello Malpighi when he placed the very thin membranes of the wings of bats and the lung tissue of frogs under the single lens of his magnifying "flea glass" and his twin-lens microscope. He was then able to see that the tiny endings of the arteries and veins (which had been seen by the naked eye for many years) were connected by even smaller, microscopic vessels (which could not be seen by the naked eye) that he called *capillaries,* which means hairlike in Latin. This development proved Harvey's theory about blood circulation correct and Galen's incorrect.

This is an excellent example of the importance of instrumentation in the advancement of medicine, as well as for all science. Neither modern science nor modern medicine would be what they are today without the invention and implementation of instruments for use in formulating and confirming our theories and hypotheses. The first magnifying instruments (microscopes) used glass beads that created distorted images. This led to the use of convex lenses placed in a tube, which led to the compound microscope. We do know that in 1609 the famous Italian scientist Galileo Galilei (1564–1642) developed the first compound microscope, using two ground-glass lenses. His design was based on a Dutch prototype.

The next big discovery concerning the circulatory system occurred in 1658 when Jan Swammerdam (1637–1680), a Dutch scientist, used an improved microscope to identify red corpuscles, found by the billions in the blood stream. This led to the knowledge of how oxygen was carried to the cells of the body. Improvements in instruments through the ages, for example, improved X-rays; the electron microscope; and computer-assisted tomography (CAT) scanning technology, which produces cross-sectional X-ray images of the body, and so on have paralleled advancements in medicine and science. Another rather new technology, MRI (magnetic resonance imaging), was originally called NMRI (nuclear magnetic resonance imaging) but the "N" was eliminated because the public held the misconception that nuclear referred to radioactivity and did not want to use it. The "N" actually refers to the oscillations of the nuclei of atoms within living cells.

Our knowledge and understanding of the circulatory system has been refined and extended but is basically unchanged since the work of Harvey, Malpighi, and other pioneers in human anatomy, physiology, and medicine.

THE MUSCULOSKELETAL AND NERVOUS SYSTEMS

Comparative Anatomy

We know that Galen (c.129–c.199) was the most influential scientist since Aristotle. Because dissection of human bodies was not permitted in ancient Rome, Galen obtained much of his knowledge about anatomy as a physician to gladiators; he observed their wounds, organs, muscles, and bones. Galen also experimented with animals, mainly pigs, which were thought to be very similar to man in their structure. He projected his knowledge of animal anatomy onto humans, leading to many of his misconceptions about anatomy. Even so, this is probably the first example of the study of comparative anatomy. In addition, Galen served as physician to several important people, including Marcus Aurelius. Based on his experiences as an experimental anatomist and physician, he wrote hundreds of books on anatomy, physiology, philosophy, and several other subjects. As reported in *The History of Medicine* (Margotta, 1996, p. 41), one of his most important books was *On the Usefulness of the Parts of the Body* in which he describes how arms, hands, legs, feet, and other organs are designed to serve a specific purpose. Much of Galen's work became gospel or medical dogma for over 1500 years, which led to the saying "knowledge was the barrier to knowledge," meaning that revered classical knowledge, writings, and documents can become a hindrance to the development of new knowledge.

Leonardo da Vinci

Leonardo da Vinci (1452–1519) studied anatomy to provide a better understanding of how to paint and sculpt the human body. He produced several writings of his studies that were never published. Had they been, more progress may have been made in the field of anatomy. It is widely reported that he dissected over 24 bodies at night in poorly lit mortuaries. From these dissections he produced many hundreds of diagrams of the muscles, blood vessels, the skeleton, nerves, and many organs. Some are still in use today.

His study of the skeletal system helped him understand the need for a site where the muscles attach, and how the muscles and bones helped form the shape of the human body. This was important for his sculpting and painting. Some of Leonardo's drawings, published after his death, show three-dimensional views of how the muscles and bones worked together, which led to his concept of the body as a machine. Along with Herophilus and Erasistratus, he is considered to be a father of anatomy.

Bones as Nonliving Structures

Ancient humans must surely have been aware of bones and muscles in the animals they killed for food. What connections they made between bones, tendons, muscles, and nerves is not known. For many years people thought that bones were nonliving structures even though evidence indicates that the bones of animals were split open so the rich gelatin and marrow found inside could be eaten. Marrow is found mostly in bones of the trunk of the body. Today we know that bone is a living, rigid, supporting tissue. The marrow found in the center of bones produces most of the red blood cells, platelets, and white cells needed for our blood supply. Bone marrow accounts for about 3% to 5% of a person's body weight. It is also very susceptible to high doses of nuclear radiation, which causes a large drop in the production of white blood cells, resulting in radiation sickness and susceptibility to other diseases.

It was not until the late nineteenth century that bones were considered to be unlike stones or rocks. Rather, they are composed of about two-thirds inorganic matter, mostly calcium and some other nonliving minerals, and one-third organic material (carbon-based living tissue). They are somewhat elastic, they grow, and they contain nerve cells. As we age, the amount of bone tissue lost through absorption by the body becomes greater than the new bone tissue formed. This causes a reduction in height and greater susceptibility to fractures and osteoporosis. We also know that most humans have 206 bones—some of us have a few more, some a few less; the ancients usually underestimated the number and importance of bones in humans. Early archeological records indicate that fractured bones were set and dislocated joints repaired. The ancient Egyptians and Jews set fractures using splints—much as we do today. To repair a dislocated shoulder they would stand the person up, place the arm over a fencelike structure, and then place a weight on the hand to stretch out the arm so the joint could come together. A very similar method is used today to correct a shoulder separation.

Trephining

Archeological evidence dating back thousands of years indicates that trephining was practiced. This procedure involved drilling a hole into the top of the skull to let out evil spirits, cure mental illness, relieve a headache, or as part of a religious ritual. This practice was common until the mid-sixteenth century. There is some evidence that it is still performed for medicinal purposes in some present-day cultures. Even today in the United States, there are a few physicians who perform this operation.

Following the basic assumptions of trephining, medical fraud was perpetrated by individuals known as "stone cutters." They claimed to cure mental illness by cutting the scalp to remove "stones" just below the skin that supposedly were causing the mental problem. They would throw a few stones (from their pockets) into a metal pail, where they would make a loud noise, claim the patient cured, and collect a fat fee. The more stones tossed into the pot, the greater the fee, and the more "cured" was the patient.

Eyes and Eyesight

As a student of anatomy in the late 1400s Leonardo da Vinci wrote about the human eye, which he referred to as "the window of the soul." He considered the eye nature's greatest work for providing understanding, with the ear next in line. Leonardo made a glass model of an eye, inserting a small hole in the rear through which he peered in order to ascertain what an eye sees. He speculated on how the image got into the eye and how the image went from the small hole in the rear of the eye, which carried the image from the optic nerve to the brain. He did not understand the nervous system, but did have a concept of the optic nerve and the role of the brain.

An argument raged over many years as to how the image of an object got to the eye. Some ancient philosophers believed that the object gave off something that then went to the eye. Plato mistakenly believed that the eye sent out a signal, which went to the object and then returned to the eye.

Euclid of Athens (c.330–260 B.C.) is best known for compiling geometry propositions and proofs from many sources and publishing them as a textbook. However, Euclid also speculated on the eye and sight. His misconception was that the eye, not the object, was the point of origin of what was viewed, and that geometry could explain how the image got into the eye.

Pythagoras (c.580–500 B.C.), a respected but eccentric Greek mathematician and philosopher founded a brotherhood or cult called "The Pythagoreans." They were dedicated to a life of speculation and contemplation of mathematics, religion, and mysticism. Later they split into two factions— the religious and the scientific. The scientific group made conceptual models of the universe and nature. They were also "atomists" who believed that everything was made of tiny particles called "atoms" and that objects gave off signals. They believed that numbers and mathematics could explain how the eye functioned. Around 500 B.C. the Pythagorean, Alemaeon of Croton (Southern Italy), was one of the first to dissect human corpses. He discovered that the optic nerves connect the eyes with the brain and that the ears' eustachian tubes connect with the pharynx (mouth). He also was the first to state that all the sense organs were connected by nerves to the brain, and that the brain was the site of human intellect. His concept was not accepted for many hundreds of years.

Galen, using his common sense, asked questions. He wanted to know, for example, how an image from a single object came to one person's eyes, yet that the same image was also available to the eyes of others. Also, how can the image of a large mountain get through such a small hole in the eye? Even though Galen described vision as related to the anatomy and physiology of the eye, the concept of seeing as an active spiritual experience involving the eye lasted up until the sixteenth century. One of the reasons for the misconceptions concerning sight and the eye was that no one could see any direct connection between the object and the eye. Therefore, there must be some spiritual or mystical explanation of "seeing" because the eye did not touch the object or have any connection to the objects viewed by the eyes.

Nervous System

The early Greeks thought that the nerves carried air or fluid, similar to the blood carried by veins and arteries. Two of the Greeks anatomists, Herophilus (c.355–280 B.C.) and Erasistratus (c.250 B.C.), who was already mentioned in the section on surgery, were interested in the human nervous system, the brain in particular. Erasistratus was the first person to make a distinction between the main part of the brain, or *cerebrum,* and the smaller part, the *cerebellum.* He also compared human and animals brains and noted that the human one had more folds and convolutions and thus more surface area than did the brains of animals. From this he correctly concluded that the superior intelligence of humans was partially due to this larger surface area of the brain.

Aristotle held the misconception that intelligence was located in the heart. Herophilus correctly placed intelligence in the brain. He also realized that the nerves were related to and responsible for sensation and motion. Herophilus made the important distinction between two types of nerves—those that receive impulses via the senses and transmit the impulses to the brain, and those that go from the brain to cause motion in the muscles and skeleton.

Not much experimental work was performed on the nervous system, possibly because the nerves were difficult to see and work with. Anatomists such as Galen, Leonardo da Vinci, and Vesalius all examined and studied muscles during their dissections. But they did not know why or how they moved the bones to which they were attached. They made no direct connections between the muscular and nervous systems.

This changed in 1766 when Albrecht van Haller (1707–1777) published his experimental work, which demonstrated that a sharp stimulus to a muscle would produce a contraction. Even more important, he demonstrated that if you stimulate a specific nerve, a specific muscle will contract. He demonstrated that the tissues themselves did not carry the stimulation, but that the nerves did. He then concluded that it was nerve stimulation that controlled muscle movement. Previously, most scientists believed that muscles just contracted when someone wanted them to move. For his work with the nervous system, Haller is considered the father of neurology.

From Haller to the present, rapid progress occurred in the field of neurology. It was determined that nerve cells are the most complex type of cell, and of all the organs/systems in the body, the brain and central nervous system are also the most complex. A German physicist named Heinrich Rudolph

Hertz (1857–1894), who is credited with the discovery of the *photoelectric effect,* also determined that the entire system of nerves is composed of many cells, each with a small gap connecting it to the next cell. He called the individual cells *neurons.* Using stains, Camillo Golgi (1844–1926) observed nerve cells under a microscope and verified these tiny gaps, which he called *synapses.* His work was followed by that of Santiago Ramòn y Cajal (1852–1934), who studied the cells of the brain and spinal column and confirmed the neuron theory. Golgi and Ramòn y Cajal jointly received the 1906 Nobel Prize for physiology.

Despite all the knowledge gained about the brain and nervous system, there are still disputes about what the mind is, the nature of consciousness, the existence and site of the soul in the body and related questions. Recent research indicates that there may be a biological basis for the mind that is located in the brain. (Historically the human soul has been thought to reside in either the liver, heart, spleen, or brain.)

DISEASES: CAUSES, TREATMENTS, AND CURES

Disease is caused by the breakdown of the body's internal environment, often as a result of external factors. As anatomy and physiology form the basis of our understanding of medicine, so is medicine the framework for the diagnosis, treatment, and prevention of disease. It might be said that the history of medicine is the history of our understanding of disease.

There are several ways to classify diseases. One is by the site of the illness or dysfunction, for example, the liver, lungs, heart, and so forth. Another is by cause, be it infectious, which means by some outside entity (organism), or noninfectious, as in an internal malfunction of human tissues, organs, or systems. This second classification can also result from something external (toxic substances) that is not infectious, or it can be caused by a breakdown in a biological process, such as cancer or aging.

Ancient Medical History

Ancient people had a simple way of classifying sicknesses. Basically, the illness was caused by an outside evil spirit that entered the body and had to be exorcised, usually by some type of priest or witch doctor. This involved superstition, magic, and incantations, and at times, serious treatments to the patient such as purging, bleeding, and trephining. Prehistoric people knew about sickness because they surely experienced many kinds of diseases. They lacked the experience, knowledge, and possibly the intelligence to address the prevention, causes, and cures of disease. So they addressed their medical problems in the best ways they knew how, often based on irrational explanations of the cause of the symptoms, and devised their own treatments based on these explanations. Natural phenomena and events influenced their ways of addressing diseases.

If something is given supernatural power, a greater power must be required to address the problem. This power was in the hands of sorcerers, priests, medicine men, and others who claimed special knowledge, particularly of the stars and heavens, and usually resulted in some radical treatment or even a human sacrifice. Ancient healers also claimed to understand, not without some rationality, the healing power of herbs and plants. Many of these ancient concepts, which are not recognized by modern medicine, are still with us today. Some examples include magic, superstition, "speaking in tongues," "laying on of hands," "ozone machines," snake-oil cures, faith healers (even by placing your hand on the screen of the TV when viewing the "healer"), and the use of crystals and icons. The National Institutes of Health (NIH) has set up a branch to provide funding for research of alternative medical practices. Some of these alternative practices address the metaphysical, religious, and psychic aspects of healing, others include practices proposed by nonscientists. They also explore unproven cures for diseases such as cancer, for example, ingesting laetrile (from apricot pits), and use of electrical stimulation to improve healing of the body.

The oldest medical records (6000 years old) were found in the Sumerian civilization of Mesopotamia. Their system was based on astrology, which determined a person's destiny by the positions of the planets and stars at the time of his or her birth. Clay tablets were used to record medical writings, including the importance of blood as the main source of all human functions.

Mesopotamia was conquered by the Assyrians and Babylonians around 2,000 B.C. The conquerors held many misconceptions about disease. They believed in a king demon who controlled lesser demons who in turn caused specific human maladies. A demon called Axaxuzu caused jaundice. Another called Asukku caused tuberculosis. Worms were thought to cause venereal diseases. The clay tablets recorded all kinds of diseases of the skin, intestines, eyes, heart, bones, and so on. Medicine was controlled by the priests, but surgery was performed by lay persons. Physicians seldom did their own surgery or dissections—these messy jobs were carried out by butchers or barbers—a practice that continued for many hundreds of years. They performed just about all types of surgery, including removing obstructions from the intestines, repairing hernias, bleeding patients for almost any physical problem, trephining, setting fractures, and removing organs thought to be diseased. Providing assistance to women giving birth was a job for the midwives. Surgeons were only involved if a caesarean operation was required.

In many ancient civilizations the physicians became specialists who were allowed to diagnose and treat a specific part of the body. Much of their trade involved magic, but they also were keen observers of nature and of how people reacted to different treatments, which helped advance the knowledge of medicine. Even so, it was important to take no chances with the evil spirits, so medicine and magic grew up together.

A record of some ancient cures and drugs have been preserved and are available for study. A mixture of frog's bile and milk was applied to infected eyes. Beer and onions were prescribed for eye relief (which would at least make the eyes water and help nature cure the problem). Animal organs, as well as animal excreta, were used to treat diseases ranging from intestinal problems to skin infections. These practices continued into the Middle Ages. Copper was used for skin and bone problems; iron for anemia and thin blood; mercury, antimony, and arsenic were prescribed for syphilis; and mercury concoctions were used to cure just about everything. Other minerals were used to treat a variety of illnesses. Treatments included taking powders, pills, spices, leaves, and roots of plants (including rhubarb, Chinese wormwood and opium, rauwolfia, and kaolin). Medical procedures included enemas, purges, bleeding, and often surgical removal of organs. In about 2000 B.C. the Egyptians had developed suppositories for use as a form of birth control.

The Greeks, with their philosophers and mythology, had a great influence on the concept of disease and its cures. Ancient Greek civilization dates back to about 3000 B.C. During these early times, Asian beliefs and medical practices were mixed with Hellenic medicine. Much of Greek medicine was based on magic, as it was in Egypt, Mesopotamia, Persia, India and other regions during these early times.

One source of information on Grecian medicine comes from Homer's *Iliad*. Much of Homer's medical knowledge was not mystical or magical, but was based on his observations of private physicians. Homer described how wounds were healed, how arrows and spears were removed, and how to use pressure to stop bleeding. He also described the use of herbs and other medicines of the day. One seemingly sensible treatment for patients was to give them wine. This prescription is used still. As Eastern mystical beliefs and practices were fused with Grecian medicine, medicine drifted more and more toward the superstitious, magical, and mythical.

In ancient Greece the word "myth" meant stories or tales. It did not matter if these tales were historical, improvised legends, or even if the tale was true or not. We usually think of Greek myths involving gods, but some involve male and female heroes, animals, and some do not involve any mythical characters at all. The Greek myths served a useful purpose in assisting the people of that time to come to some theological and common-sense understanding of the Earth, the heavens, the environment, and reasons for being ill.

We will now consider some of the beliefs and misconceptions about infections, diseases, and their cures that have existed from ancient times through the twentieth century.

Later Ages of Medicine

After the decline of the Roman Empire, many civilizations in Europe, the Mediterranean region, and Egypt experienced the Dark Ages, a period that lasted from the Greek/Roman classical age (from the fall of Rome in 476 A.D.) until the Renaissance, which began in Italy in the fourteenth century. The term "Dark Ages" as well as the historical period it connotes is particularly relevant to medicine and science because many records were lost and new knowledge was not generated. We are fortunate that the scholars of the Muslim and Eastern countries salvaged and added to much of the writing and knowledge that was destroyed during the Dark Ages.

It was not until the nineteenth century that physicians had a concept of disease as we know it. Over the centuries it was believed that illnesses were caused by a disruption of the body's four "humors," which consisted of the sanguine, choleric, melancholic, and phlegmatic, as well as the three bodily fluids: arterial blood, venous blood, and nerve fluid. All of these were thought to be controlled by human and animal spirits. Therefore, cures could not completely ignore the "spirits."

Diagnosis

The diagnostic techniques that predate the nineteenth century consisted of observing the color, thickness, and smell of the blood, urine, sputum, and feces. Granted, these types of observations are of some importance today, but how these observations are used to diagnose diseases is much different. For instance, in the past if the urine was cloudy at the top of the bottle, the cause of the illness was in the head. If there was a deposit at the bottom of the bottle, the cause was in the lower part of the body. If the blood was too thick or thin, the common treatment was to bleed the patient. Treatments were based on the old Galen "theory of opposites," which meant if a pain was on one side of the body, you bled the patient from the other side.

Bleeding

Bleeding existed as a common treatment for many illnesses over many centuries. It was believed that bleeding "thinned" the blood and cleansed the body of bad spirits and other ill effects, in addition to reducing blood pressure. Bleeding was rather popular during the Middle Ages. It was accomplished either by cutting a vein and letting blood leak out, or by attaching blood-sucking leeches to the patient. Another popular method used a cupping vessel, in which a small piece of cloth was burned inside the cup to create a vacuum, which then sucked out blood from a small incision. Patients could attend public baths where an attendant would perform the bloodletting. As an alternative, the patient could go to a barber/surgeon who would consult astrological charts to figure out the best days for the patient to be bled.

Alchemy

Alchemy is one of the most interesting ages of medicine. One of its practitioners, Paracelsus (1493–1541), whose real name was Philippus Aureolus Theophrastus Bombastus von Hohenheim, is one of the most colorful characters in medical history. There is some dispute as to exactly when or where alchemy began. It is safe to say that its roots began about 3,000 years ago with the development of a philosophical outlook on illnesses and the use of drugs. The word *alchemy* comes from the combined Arab words, "al" for an object or thing, and "kimia," which is thought to mean chem or "khem," an ancient word for Egypt. Or the origin may be from the Greek word *chyma,* which means to melt and cast metal. The use of chemicals or drugs to cure illnesses was developed simultaneously in China, India, Syria, Persia, Greece, Palestine, Egypt, Alexandria, and perhaps other countries, including Africa. It was not until the twelfth century that the knowledge of alchemy was transported west by the Islamic countries to the Latin countries, Spain, and western Europe. The first Latin translation of an alchemy text from Arabic was the *Book of the Composition of Alchemy,* published in 1144. The age of alchemy refers to the period when medicine and chemistry were practiced in Europe during the Middle Ages and early Renaissance. It is now believed that most physicians/ alchemists were cheats in the sense that they used magic, astrology, and superstitions to form a general prognosis for patients with all kinds of illnesses.

Alchemy developed two major principles: First, that all matter was one, having a common soul or unity, which led to the use of commonly known elements and metals; second, that all matter could be "transmuted" from one form to another by the use of a *philosophers' stone,* which grew out of the philosophical/religious aspects of alchemy. Together, these components led alchemists in several directions. One was the search for the philosophers' stone, which was needed to transmute base metals such as lead into gold. Their other goal was to use the philosophers' stone to prepare the perfect medicine for people, called the *elixir vitæ* or elixir of life. In her book *The Scientific Renaissance, 1450–1630,* Marie Boas Hall (1994) sums up the attitude of alchemists as follows:

Most alchemists shared Tycho's [Brahe, 1546–1601] arrogant certainty that only the alchemist could judge whom to initiate into the "subtle science of holy alchemy," and that only initiates could be trusted. Certainly they were successful in keeping their meaning secret, and few who are not alchemists can pretend to understand what they wrote on the subject. (p. 172)

Because Aristotle taught that there is a hierarchy in nature, that gold is the most noble of metals and man is the most noble of living beings, then the philosophers' stone and elixir vitæ were considered to be the natural means of achieving perfection for both the inanimate and animate worlds.

Over the centuries many alchemists claimed to have discovered either the philosophers' stone or elixir vitæ. They thought the ore, cinnabar, which yields liquid mercury when heated, or sulfur compounds (brimstone) that produce sulfur fumes when heated, as well as many other substances, singularly or in combination, were the long sought after stone or elixir. Many patients died from mercury or other types of metallic poison that were prescribed by their Doctors of Physics, as some alchemists were called.

Even though their beliefs and misconceptions about transmutations of elements were incorrect, they weren't all that wrong. In recent decades one element has been changed into two or more different elements or elements were combined to form other elements through the processes of nuclear fission and nuclear fusion. Some of the alchemists' concepts of using drugs to combat disease have now become a mainstay of modern medical practices. Alchemists were very confused and not very honest about the particulars, however.

Drugs

The use of drugs to cure human illnesses and relieve suffering goes back as far as known history. Pharmacy is an ancient profession whose practitioners include medicine men or witch doctors and women of the tribes who collected plants, roots, and herbs for use in treating illnesses. We still use some of the thousands of plants that have been identified as having some medicinal value, or we artificially produce some of the same chemicals that these plants produce. Today, large computer databases have been developed to record the numerous plants with potential medicinal benefits.

Many drugs extracted from plants are very toxic, but when used in correct amounts they can be effective. For example, the bark of the cinchona tree is used as the source of antimalarial quinine. Digitalis, which comes from the leaves of the foxglove plant, is used as a heart stimulant. Nightshade is the source of belladonna, needed for antispasmodic atropine, which is used to treat eye problems. The plant rauwolfia is used to produce tranquilizers. The plant guaiac ("holy wood") was brought back from America to treat the "French disease" known as syphilis. It was thought to be effective, but eventually proved useless.

A Greek physician, Pedanius Dioscorides (c.40–90), classified about 600 medicinal plants and thousands of drugs. His classification was alphabetical and many of his drawings and descriptions were inaccurate. Even so, Dioscorides may be considered the father of *pharmacology,* which is a Greek word meaning "the study of drugs."

In the first century Dioscorides wrote an important book entitled *De Materia Medica.* He was the first to describe opium and what was later known as laudanum, an extract of alcohol and opium plant blooms. Another important source of information for drugs used in medicine was the twenty-eight volumes written by Zosimus of Panopolis (c.300 A.D.). These texts described the extraction of mercury from cinnabar, how to obtain white lead from litharge and vinegar, and how to extract arsenic from realgar. Zosimus also described how to use verdigris (copper sulfate), antimony, ocher, natron (sodium), sulfur, iron, silver, gold, mercury, and other substances to treat a great variety of illnesses, including mental illness, gastric distress, arthritis, venereal diseases, consumption (tuberculosis), blood disorders, kidney malfunction, constipation, plagues, and so on. The use of drugs has always been an important aspect of the practice of medicine. However, most of our knowledge came from trial and error and by observing the results once a drug was used to treat a specific disease. It was not until the nineteenth century that pharmacy was based on scientific knowledge of how specific drugs interact with specific diseases and the human body.

Two branches or divisions of physicians emerged based on their beliefs on the use of drugs to treat illnesses. One group was called the Paracelsists ("spagyrists"), or those who believed in medical chemistry, also known as iatrochemistry. The other group comprised the herbalists who believed in the use of plants, roots, herbs, berries, and so forth.

Paracelsus

At the beginning of this section on alchemy, an individual was mentioned who had an impact on medicine. Theophrastus Philippus Aureolus Bombastus von Hohenheim always addressed himself as "Paracelsus," supposedly after the ancient Roman medical philosopher Celsus. He added the "para" because he thought he was not only equal to Celsus, but above him. Paracelsus was the son of a Swiss physician, who was initiated into alchemy early in life. Paracelsus was interested in astrology, religion, and the use of minerals as medications to achieve his main goals, which were to reform medicine and heal mankind. Paracelsus is credited with introducing *iatrochemistry,* which is the use of chemicals for medical purposes. He also believed that the chemical (material) and spiritual (religious) ingredients of alchemy were united. His views irritated his opponents because he not only questioned their medical practices but also their motives. Paracelsus challenged the Doctors of Physicks to heal as many patients as did the unschooled peasants who used old folk medicines and techniques. He greatly annoyed other practitioners of medicine when he challenged their efforts to establish Galen as the final authority in medicine.

Although Paracelsus was good at experimenting with different drugs for treating different diseases he held some erroneous ideas. For instance, he believed that the human body was essentially a system of chemicals composed of three principles: mercury, sulfur, and salt. Illness was caused by an imbalance of these three principles and the illness could be cured by restoring the balance. He did get one treatment right. He treated anemic patients with "poor blood" with iron salts—but his rationale was that the iron salts were red, just like the planet Mars and the blood.

Paracelsus's challenge of Galen is sometimes referred to as the first Scientific Revolution. It predates the challenge of Ptolemy's earth-centered solar system by Copernicus. As the story goes, Paracelsus tossed a copy of Galen's writings into a bonfire to show he was serious about reforming medicine. He even expected his beliefs to replace the Hippocratic Oath, but he upheld the beliefs that a doctor's character was of utmost importance and his place was beside the patient. Paracelsus's mystical beliefs related to astrology, but this did not prevent him from saying that his writings and teaching would be based on his observations and experiences with patients. His arguments were usually by analogy and based on what he could observe by trying things out. His epistemology of *experienz,* as he called experimenting, led to the development of medicine and the concepts of the scientific method later developed by Francis Bacon and others in the seventeenth century and beyond.

Paracelsus's big opportunity came when he was called to treat Johann Froben (also called Frobenius) in 1526 for a critical illness. Froben had a serious leg problem (probably phlebitis) and surgeons wanted to remove his leg. Paracelsus was called and cured the problem with one of his secret chemical mixtures. It just happened that Erasmus (1466–1536), who was living with Froben at the time, was also ailing. Paracelsus cured Erasmus's digestive problems, probably with laudanum. In gratitude, they secured Paracelsus a post as town physician and an appointment to the local university. Because of his ideas, his refusal to take the Hippocratic Oath, his lack of a medical degree, and the fact that he taught his students in vernacular German instead of Latin, he was ostracized by the faculty at the university.

When Froben died, Paracelsus lost his major source of support. Doctors, students, and the entire medical profession were against him. One reason for their enmity was that he kept attacking them for the large fees they charged, even when their patients did not recover and died. A high church official promised Paracelsus a large fee if his intestinal problems could be cured. He cured the official with opium that he converted into laudanum pills, but the official refused to pay. Paracelsus sued and lost his case. He was destitute, denounced the judge, and was forced to flee for his life.

Paracelsus never held a job in a university or any other organization again; he was a medical pariah. He then moved to Nuremberg where he stirred up more trouble in the medical and religious communities by attacking the way they treated diseases. In particular, he objected to their treatment of syphilis with "holy wood"; he treated the disease with mercury concoctions. He was prevented from publishing his writings, which forced him to finally go back to the mines of Tyrol to study miner's diseases. In his book entitled *On the Miners' Sickness and Other Miners' Diseases* (1567) he showed that miners' sickness was a disease of the lungs and that the dust they breathed also caused ulcers and skin diseases. He identified the symptoms of antimony, arsenic, mercury, and alkali poisoning men contracted in the mines. He might be called the father of occupational medicine. Paracelsus died at the age of 48 and was buried in a house for the poor.

It was not until after Paracelsus's death that his writings were published and he achieved the recognition he desired. His influence continued until the death of alchemy in the middle of the seventeenth century. Every age needs iconoclasts like Paracelsus who use their uncommon sense when approaching the barriers to knowledge. The very few iconoclasts can sometimes correct our mistaken scientific beliefs and misconceptions.

Causes of Infection and Diseases

The cause of infection and diseases was attributed to any number of incorrect beliefs. Most of these misconceptions were related to spirits, demons, witches, an imbalance of the humors, bugs, worms, snakes, or some other explanation that seemed to make sense, but for which there was no evidence.

Specific diseases were identified, but their causes and effective cures were unknown. An example can be found in the ancient disease syphilis. Syphilis became an epidemic in Europe in the late fourteenth century as it was spread by armies invading one country after another. One theory was that it was caused by a worm. Oddly, it was not known how the disease spread until the Scots became aware in the early 1400s that a person could be infected by sexual contact. This enabled people to take prophylactic and preventive measures, but this did not seem to reduce its incidence. There is some evidence that a particularly virulent mutation of the spirochete bacillus increased its rapid spread to people who may have been previously immune to a milder form of the disease. Syphilis spread to Asia, Africa, and America as sailors visited these areas. Many ineffective treatments were tried prior to the development of penicillin in 1941.

Girolamo Fracastoro (1478–1553) theorized that an invisible germ caused disease. He distinguished live contagions that caused disease from poisons and other human maladies. He thought that by using some poisons or "caustics," one could kill the contagion or germs. He classified contagion as occurring (a) by personal contact, (b) through contact of one's clothes with another person, or (c) by transmission through the air. The development of both cell and germ theory had to wait until the development of the microscope in the late 1500s. The development of medical instrumentation was crucial for the correction of many misunderstandings in medicine.

In the early centuries keen observers must surely have noticed that a dew-drop on a leaf magnified the surface of a leaf. Someone with uncommon sense could translate this phenomenon and then prepare a convex glass similar in shape to the drop of water. Lens makers did just this to prepare corrective lens for eyeglasses. The discovery of the microscope is attributed to at least two people. The Dutch spectacle makers Hans Janssen and his son Zacharias (c.1580–1638) made lenses that enlarged images and corrected poor vision. Zacharias placed a convex lens at each end of a tube and noticed that it enlarged the image of small objects. This is considered to be the first compound microscope. Later credit was also give to Galileo for improving the compound microscope. At about the same time, Hans Lippershey (1570–1619) also developed magnifying lenses, but it is not known if he combined them in a tube as did Zacharias Janssen. There is also some evidence that one of the first microscopes was comprised of a small glass bead imbedded in a hole in a piece of wood, or at the end of a tube, but the definition of the image was so poor that it was of no use.

Cells and Germs

In 1665 the Englishman Robert Hooke (1635–1703) was examining cork under his microscope when he saw little divisions between sections of the cork. He thought they resembled the small rooms in monasteries know as cells. This was the first time anyone identified anatomical structures smaller than tissues, muscles, or nerves.

It was in 1673 that Anton van Leeuwenhoek (1632–1723) developed a more powerful compound microscope. This invention ended forever the misconceptions concerning diseases, germs, infections, and so on that persisted from the beginning of time through the age of alchemy. Later, Edward Jenner (1749–1823) anticipated the germ theory when he realized that people who had recovered from the dreaded smallpox did not get the disease again. He correctly surmised that they developed something that prevented the disease from reoccurring. He experimented by scratching or inoculating an 8-year old boy with the milder cowpox. After 6 weeks when the boy recovered from the cowpox, Jenner inoculated him with smallpox, which the boy did not develop. He published his results in 1798 and received much public criticism, which persists today, for his use of a human experimental subject without knowledgeable consent by the subject.

After 1831 things developed rapidly when Robert Brown discovered the nucleus of plant cells. Then in 1835 Jan Evangelista Purkinje (1787–1869) made the discovery that the skin of animals also contains cells, whose walls were thinner and not as well defined as those observed in plant cells. Unfortunately, no one paid much attention to these discoveries until a few years later. Matthias Jakob Schleiden (1804–1881) and Theodor Schwann (1810–1882) fully developed the cell theory for plants and animals. They both received credit for their cell theory, which states that all living things are composed of cells and that all cells arise only from other cells. The cell theory (cytology) is the basis for modern medicine and biology.

The cell theory led to the germ theory, which holds that infectious diseases are caused by microorganisms within the body. The concept that all cells arise from other cells, and that diseased cells (and tissue) come from healthy cells was the basis for what is now known as cell pathology. This was radical thinking, because it refuted the common misconception of spontaneous generation, which purported that living matter and diseases come from nonliving matter.

Louis Pasteur (1822–1895) is credited as the father of the germ theory. Although his name (pasteurized milk, etc.) is well-known today, his germ theory did not gain immediate acceptance in the scientific community. Much of Pasteur's work was based on microscopic examination of cells and germs. Pasteur and others found and identified specific microorganisms that caused specific diseases; for example the rod-shaped bacteria (tubercle bacillus) that causes tuberculosis, the ball-shaped bacteria (gonococcus) that causes gonorrhea, and the corkscrew-shaped bacteria (spirochete) that causes syphilis. He also conducted experiments that prevented diseases and found specific cures. For example, Pasteur discovered that when growing cultures of some bacteria, they could be killed by placing other organisms in the same culture. This led to the later development of penicillin and other antibiotics. He also developed a method to kill diseases present in cow's milk (pasteurization) by heating the milk to a point below boiling (about 65°C for 30 minutes), which would kill harmful microorganisms, but not spoil the milk. In addition, he finally put to rest the mistaken belief in spontaneous generation.

Septic and Antiseptic

In 1865 Joseph Lister (1827–1912) discovered antiseptics (a substance that kills or prevents the growth of germs), which as a surgeon he used to reduce infection in operating rooms. Again, the development of new knowledge that supplanted common misconceptions resulted in advancements in medicine. It was some years before his concepts of septic (from the word *septikos,* which means "rotten" in Greek), and antiseptic (which means the destruction of microorganisms that cause diseases) were accepted.

Ancient physicians had no concept of *septic* or *antiseptic.* Even before the germ theory was well known, the Hungarian obstetrician Ignaz Philipp Semmelweis (1818–1865) questioned why so many women died in the maternity wards of what was called "invisible vapour." He noticed that women giving birth who were attended by surgeons who had just completed postmortem autopsies had a very high death rate. In 1847 he instituted the practice of cleaning hands, knives, equipment, clothing, and so on before treating patients. The death rate dropped but he did not know why; he surmised that there was some connection between the dirty hands and equipment and the deaths. It was the American surgeon William Stewart Halsted (1852–1922) who instituted the practice of nurses and physicians wearing rubber gloves—it was easier to sterilize the gloves than the hands.

Epidemics and Plagues

Throughout history great epidemics and plagues claimed many millions of lives. The causes were not known, and the knowledge for treating these diseases was nil. The only common circumstance was the high death rate that occurred once an epidemic spread, usually beginning in the east and moving westward. Rats were known to carry the "black death," which claimed over 25 million lives. Although they tried to care for the sick by using all the alchemy, chemicals, herbs, and metaphysics at their disposal, doctors were powerless against this plague. They were overwhelmed. Ships were prevented from landing in harbors, refugees were turned away from cities, and still the epidemic spread from country to country. Some say epidemics may be one cause for the decline of the Roman Empire and the beginning of the Dark Ages in Europe. The plagues in Europe directed people away from rational approaches to problems. Because of medicine's inability to cope with these problems, Christianity and other religions flourished. Exorcisms were performed out of the belief that devils caused the epidemics. The Christians of the day did much to relieve suffering. They established hostels and ran hospitals, and they comforted the sick. No one knew what to do; the epidemics took their toll and ran their course. Many superstitions were developed to ward off or cure the plague. Since the days of the Greek gods it was thought that the laurel tree was a healing plant that provided protection against evil. During the plague years it was considered fortunate to have one of these trees growing outside your home. People carried leaves from the tree in their pockets and kept them under their pillows to protect them from contagion and evil spirits—but to little avail. How many lives could have been saved if what we know now about diseases, their causes, treatments, and cures could have been applied in those dark days?

Superstitions

Superstitions related to causes and cures of diseases go back thousands of years. Some are based on prehistoric folktales, others on myths, and still others are based on more recent beliefs from many countries, including old England, Europe and the Americas. *The Encyclopedia of Superstitions* (Hole and Horsley, 1996) lists over 200 superstitions that tell how we get, cure, and prevent diseases. The following are some of the 81 categories of superstitions listed: (1) consumption; (2) dysentery; (3) agues and fevers; (4) gout; (5) heart disease; (6) hysteria; (7) influenza; (8) pneumonia; (9) scarlet fever; and (10) whooping cough. We list just a few examples below.

In Scotland it was generally believed that butter made from the milk of cows who had grazed in the churchyard would cure consumption (tuberculosis). In England ashes of burnt human bones found in churchyards were mixed with ale to induce intoxication. This concoction was used to cure dysentery. Skulls and bones are used in many cultures to fulfill superstitious beliefs. In Africa the practice of "throwing the bones" is still used in various ceremonies. In Australia a "singing magic" is sounded into a bone pointed at the victim, who then becomes ill and dies. Those with headaches can drive a nail into a dead person's skull to relieve their own pain.

The leaves of aspen and poplar trees move with the slightest breeze, and are therefore referred to as "shivver-trees." As "likes were thought to cure likes," the leaves of these trees were used to treat agues and fevers. There are as many superstitious cures for agues and fevers as there are testimonials from surviving patients.

An Irish cure for influenza was to make a plaster of the clay found on one's doorstep and lay it on the patient's chest. Both the clay (earth) and the threshold were considered sacred, visitors reenforced this concept by saying "God save all herein" before they entered to visit the sick.

To cure a child of whooping-cough you placed some of that child's hair in the notches of the bark of a special tree (usually an ash). Warts could also be removed by sticking pins in the tree, then in the warts, and again back in the tree, where the warts will be transferred. Another wart cure prescribes the sufferer to cut a potato in half, rub both halves on the wart, bury the potato, and when it rots the warts will be gone.

Our final example is of a widespread superstition thought to be the cause of many human ailments. The *evil eye* is a superstition that dates back to the Egyptians and Greeks. In some cultures it was thought that a person who was cross-eyed or who had some other physical imperfection would cause either illness or misfortune if they looked at you. Or, if someone walking with his or her head lowered and eyes on the ground just happened to look up at you, you were cursed. Even today in the lower Rio Grande Valley area of the United States some people believe that their children became ill because someone gave them the "look," as they call it. This caused illness in the child or resulted in some other misfortune.

Faith Healing

Faith healing has a special place in medicine. It is usually considered a process that restores a person's mental, spiritual, or physical health through the intervention of divine spirits. It is not the same as modern medicine, which is based on science, or prehistoric medicine, which was based on folk tales, herbs, and so on. Faith healing was used by the Greeks and was later incorporated into Christianity. The orthodox doctrine prescribes faith as a prerequisite to healing and health. At times it involves not only prayer, but laying on of hands, speaking in tongues, and confession. Some religions accept a combination of faith and modern medicine for the well-being of the faithful, whereas others practice rituals that they believe will enable the divine Spirit to overcome the evil spirits. See the section below entitled Alternative Medicine and Modern Medical Practices for the role of the placebo effect in faith healing.

HEALTH AND PREVENTIVE MEDICINE

History of Preventive Medicine

For the most part, prehistoric people were rational. Their existence depended on surviving the elements and securing food. More important, they learned from their environment and, in time, used fire and other things found in nature to their advantage. There is some evidence that ancient civilizations had some conception of illness and cleanliness, both individually and communally.

The following is just speculation because little is known about the health practices of ancient civilizations. Early people were nomads. They moved where the food supply was most available. No doubt, they disposed of garbage and human feces wherever they happened to be at the time and whenever they felt the need to do so. This probably caused no health problems, as they just moved on to another location. It was not until about 12,000 B.C. that man domesticated animals; agriculture was developed 4,000 years later, around 8,000 B.C. These advancements favored the development of more permanent communities. This must have caused some problems relating to the disposal of garbage and human waste. One can speculate that they did what modern armies still do when they settle down. They dug slit trenches or waste pits, giving the job to the person of low status. With the establishment of permanent communities, tribes did not just flee the approach of hostile tribes as they previously had; rather, they stayed to fight for their villages, food, and clothing. This required a water source and an advantageous elevation on which to build their communities and erect barriers to outsiders. This type of settlement brought with it problems related to public health and personal hygiene, as well as the need for preventive medicine.

About 5,000 years ago cities in India developed sewage drains and public baths. Personal hygiene became part of their culture. They followed codes of self-discipline, practicing food and environmental sanitation. At about the same time they developed heath education and health care, which included the use of medicines.

Babylonian ruins, dating before 2,000 B.C., contain large stone drains and privy pits that most likely were used for sewage purposes. Similar stone drains were directed into the Nile river in Egypt. These seem to be the earliest evidence that people living in permanent settlements consciously made efforts to promote public health. Even though cities were built near water, there was always a need for clean water. Some wells were dug outside the cities, and some canals were developed. Artificial masonry *aqueducts,* a Latin word meaning a drawing off of water, were constructed in 704 B.C. in Assyria to bring water to the city of Nineveh. In 715 B.C. aqueducts were built in Judah to bring water to the city of Jerusalem. The most famous aqueducts are those built in Rome during the Roman Empire. About 2,500 years ago the Romans built what was called the Cloaca Maxima (large opening), which was a large drain designed to dry up a swamp, but which was also used as a sewer system that drained into the Tiber river. An exceptional aqueduct that is still in use was built to bring water to the city of Segovia, Spain. The water is carried in a structure composed of arches of stone that reach 100 feet above ground. Many of these old aqueducts also had reservoirs used to store water during droughts.

Some of the Asian countries had a different solution for disposing of human waste. Each home had a small area with a small door that opened to the outside where the excreta was deposited. Early each day the collectors emptied the "honey pots'" contents in fields, where it was used as fertilizer.

Garbage and Waste Disposal

As cities grew, so did the problems associated with garbage and waste disposal. In the eighteenth century, city dwellers would toss both garbage and human waste into the streets, waiting for a crew to come along and clean it up. The garbage accumulated in piles and eventually the piles were dumped in rivers, contaminating the drinking water. Private privies were built until it was determined that a concentration of privies contaminated the shallow wells used for drinking water. During the Middle Ages the rivers in England and Europe became open sewers.

By the seventeenth century physicians became aware of the connection between disease and the disposal of human wastes. The general public remained unenlightened until the nineteenth century, when waste disposal become the responsibility of local government in larger Western cities.

Even before the causes of diseases such as cholera were known, physicians made the connection between filth and disease despite the fact they did not understand the exact cause and effect at work. During the plagues and various epidemics of Europe and England, physicians realized that contagions were linked with the occurrence of devastating diseases. They insisted that the dead be burned to help stop the plague. They recognized that rats were somehow involved, but complete elimination of all rats was impossible (just as it still is today in any large city. It is estimated that there are more rats living in New York city than there are people). The history of plague epidemics in Western Europe covers several centuries. The first plague came from the east in the year 542. The second plague, called the Black Death, arrived in England in 1348 and lasted more than 50 years. It ultimately killed approximately half of the European population. The third came from China, by way of Hong Kong, in 1849 and was experienced worldwide. Partly because of public health awareness, this plague's death toll was not as great in the United States as it was elsewhere. We still have occurrences of viral infection today, but they are not nearly as severe as even the "influenza" outbreaks of the more recent past, such as the one that occurred during World War I that killed hundreds of thousands of people. Influenza derived its name from the words *influentia coeli,* an old Latin term for the disease dating from the Dark Ages, when people believed the sickness was caused by some "influence in the sky."

During epidemics and plagues physicians did what they could for patients. They instigated the cleanup of unsanitary areas and restricted travel. Health officials quarantined homes and patients. To limit contagion patients with diseases such as diphtheria, scarlet fever, smallpox, and tuberculosis were once quarantined. With the advent of modern public health quarantines are seldom used.

Personal Hygiene

Prehistoric people had a difficult time keeping themselves clean. Their only clothing was made of the hides and fur of animals. If they lived near a river or lake they could bathe in warm weather, but as they moved north bathing became more difficult. It is not at all clear just how much attention was paid to personal hygiene throughout history.

Over 3,000 years ago both outdoor and indoor communal bathing was instituted in Greece and Rome. For many years these public baths were used not only for bathing but also for feasting, drinking, and as a form of therapy (including bloodletting). Mixed-gender public bathing was popular in several countries, including Germany, Switzerland, and Japan, as well as in Greece and Rome. These ancient baths were the "community centers" of their day. The practice of public bathing died out as debauchery increased and knowledge of how diseases are spread grew.

Personal hygiene and bathing assumed a religious meaning in some cultures, in both old and currently practiced religions. Many years ago in Germany, *shvitzbud,* or Turkish "sweat baths," were used by male Jews to cleanse themselves for their Sabbath. They were not allowed to bathe in the streams and rivers because they might contaminate the water for Christians. Several religions in India require daily bathing. Even today, large water troughs are distributed throughout cities where people gather to bathe as best they can—while still clothed.

Public Health

The concept of public health developed slowly in various countries over the ages. For example, the importance of clean, fresh air was not considered until the Industrial Revolution. The air in England during the Industrial Revolution was far more polluted with soot and smoke than it has ever been in any city in the United States.

Sir Edwin Chadwick (1800–1890) is considered one of the fathers of public health for his insistence on cleaning up the towns of England. He demonstrated a positive correlation between slums and high mortality. In 1848 Chadwick's efforts were responsible for passage of the first Public Health Act in England, which made the state responsible for the health of the public. Shallow wells were abolished. Chadwick also designed ceramic pipes to carry sewage, thereby reducing drinking-water contamination.

John Snow (1813–1858) collected data demonstrating that cholera was a water-borne disease. His data was collected from one particularly densely populated block in London. He traced the infection to drinking water contaminated by a leaking cesspool. Although he did not know the exact cause of the disease, his statistical approach led to an emphasis on cleaning up the physical environment rather than just relying on medical treatment of individuals. Snow might be considered the father of epidemiological investigation because he used statistical data to arrive at his conclusions.

Preventive Medicine

It was soon realized that preventive medicine, which was seldom practiced, was not the entire answer to a healthy population. Governmental efforts to improve the physical environment of cities and reduce poverty continued. The city fathers realized that poverty and unhealthy workers do not generate taxes, besides which it cost money to care for those who were sick and impoverished.

Preventive medicine has a varied history, including its current place in modern medicine. As previously mentioned, ancient people tried to prevent illness by performing all kinds of rituals, both spiritual and physical. Over the past several thousand years people ate certain foods, submitted to dangerous surgery, bled themselves, took dangerous drugs, and engaged in other physical practices to prevent as well as cure illnesses.

It is significant that after World War I there were no major plagues or disease epidemics. This led to a surge in population growth and an increase in the proportion of older people in Europe. From 1830 to 1930 over 50 million people emigrated from Europe, mostly to the Americas. These demographics are considered partly the result of the practice of public health and preventive medicine, as well as improvements in drugs and medical practices.

Diet and Nutrition

Early people did not pay much attention to their diets. They were busy getting enough to eat and staying alive. They were aware of the benefits of different kinds of food, however, and they obviously ate meat as well as fruits and vegetables. *Homo sapiens* is one of the few species of mammal that is both omnivorous (eats both plants and animals).

The science of human nutrition deals with the nutrients in foods and how our bodies assimilate them. Today we know a great deal more about the major food groups and the benefits of vitamins and a well-balanced diet than people did just 50 years ago. The major food groups are: (a) carbohydrates, which should make up about 55–60% of one's daily caloric intake; (b) fats, which should make up no more than about 30% of your diet; (c) proteins (amino acids needed to build tissue), which should comprise at least 10–15% of one's daily caloric intake; and (d) fiber, which is not actually a food group, but is very important. Fiber from the cell walls of fruits, grains, and vegetables is not broken down in the digestive track as are the foods

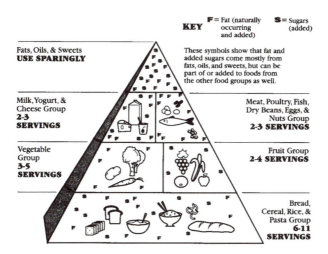

Figure 2.3 The food pyramid used as a daily guide for a balanced diet. Foods located at the top of the pyramid are eaten sparingly, whereas those at the bottom comprise most of a balanced diet. SOURCE: U.S. DEPTS. AGRICULTURE AND HEALTH AND HUMAN SERVICES.

from the other three groups, but fiber is essential to good health and reducing constipation. Early Americans did a lot of physical labor, which required a lot of calories and high-energy foods. This meant a diet high in meat and fat. As the American lifestyle became more sedentary, our dietary requirements changed, and a reduction in proteins and fats is now recommended. Unfortunately, the misconception exists today that one should not eat meat, eggs, dairy products, or fats. A major misconception is that one should strive for a diet without any fat, meat protein, cholesterol, or salt. First of all, the body cannot long survive without fat and/or protein. Second, some cholesterol, of both types, is a requirement for a healthy body. Third, there never has been and probably never will be a diet absent of all salt. The problem with the American diet today is that we do not need these foods in the same proportions as did our forefathers, but these foods are important for a healthy nutritious diet and should not be eliminated totally.

The so-called *food pyramid* (Figure 2.3) is a good guide for a balanced diet. It recommends about two or three servings daily of poultry, fish, lean beef or pork, and no more than three eggs; four or more daily servings of green and yellow vegetables and fruits, those high in vitamins A and C, such as broccoli, tomatoes, cantaloupes, carrots, green beans, and "greens" of all kinds; six or more daily servings of carbohydrates such as whole-wheat bread, pasta, and cereal; two daily servings of dairy products such as low-fat milk and cheese; and two or more servings daily of polyunsaturated oils such as corn, safflower, soybean, or peanut oil, (but not tropical oils such as coconut or palm). Depending on one's age and level of physical activity, an adequate diet including all food groups is required for a healthy body.

Vitamins

There are two general classifications of vitamins: fat soluble (A, D, E, and K) and water soluble (C and all the B vitamins). Water-soluble vitamins are not stored in the body because they are quickly excreted in the urine. Consuming an excess of supplemental vitamins (megadoses) may cause more harm than good, and is not necessary if a healthy diet is maintained. This is particularly true for fat-soluble vitamins.

The history and misconceptions of some diet deficiencies are instructive. The Portuguese navigator Vasco da Gama (1460–1524) voyaged to the Cape of Good Hope in 1497, which took many months. Many of the crew became disabled and died because of a dietary deficiency. At the same time Great Britain was expanding its navy and merchant fleet and needed to solve the problem of illness among the crew members. The naval physician James Lind (1716–1794) realized that the food carried on board for long voyages was either salted or dried to keep it from spoiling. This meant a diet without fresh fruits or vegetables. Lind knew that for hundreds of years Dutch sailors always carried citrus fruits on their voyages. He experimented and found that citrus fruit prevented scurvy. It took the British Navy another 50 years to implement a policy of providing fruits for long voyages. At first they supplied lime juice (thus the slang term "Limey" for British sailors); later other citrus fruits that were higher in vitamin C (ascorbic acid) were carried on board or searched for on voyages. Unfortunately, before Lind convinced the British Navy to supply citrus for its sailors many suffered from bleeding gums, bruising, and wounds that would not heal. Some never recovered and died of scurvy or its complications, particularly if they were on long voyages.

Christiaan Eijkman (1858–1930) went to the Dutch East Indies to study beriberi, a disease that causes a wasting of the muscles, which can cause paralysis and death. Eijkman was familiar with Lind's work with citrus and vitamin C, so he figured there was something similar missing in the diets of these island people. He found that when their diet consisted mainly of polished rice, in which the hulls were removed, they would get beriberi. Once whole-grain cereals were introduced the problem was solved. Later it was determined that vitamin B_1 found in the hulls of grains prevented beriberi. Prior to this discovery, people who did not eat whole-grain cereals got beriberi and suffered from edema, stunted growth, wasted tissues, confused mental status, low morale, and eventually died.

Pellagra is another disease caused by a dietary deficiency—the lack of complex vitamin B. It causes weakness, insomnia, weight loss, rough skin, and painful mouth sores. It is not common in the developed world, where whole-grain cereals are eaten and niacin and other B-complex vitamins (a combination of B vitamins) are added to foods.

Rickets is a disease that deforms the skeleton, which is caused by a lack of vitamin D. Vitamin D is required to maintain proper levels of calcium and phosphorus in both bones and cartilage. Vitamin D is absorbed from exposure to sunlight and is usually added to milk and some other foods. It must constantly be ingested because the human body does not manufacture vitamin D.

Adequate vitamin A is easily obtained in a healthy diet that includes green/ yellow/orange/red fruits and vegetables. At one time a lack of vitamin A caused night blindness, so we were all told to eat our carrots. Vitamin A is important for healthy skin, bones, teeth, and reproduction, as well as for vision. In children, vitamin-A deficiency can cause blindness. Some people have the misconception that high levels of vitamin A are beneficial; on the contrary, ingesting an excess of vitamin A can poison you.

One of the most persistent misconceptions is that vitamins are an adequate substitute for a balanced diet. This misconception began about 50 years ago when more knowledge became available about the role of vitamins in good nutrition. The only persons benefiting from the excessive use of vitamin supplements in the United States are the ones selling them. Our bodies pass off in urine and feces most of the excess water-soluble vitamins we ingest. Our bodies manufacture some vitamin D via exposure to sunlight, and we obtain vitamin K from fats. The other vitamins required for a healthy body can be obtained by eating a balanced diet. Taking in an excess of fat-soluble vitamins (A, D, E, and K) may produce toxic symptoms.

Physical Fitness

Physical fitness is required for preventing illness and maintaining health. Not many of us today do 10 to 12 hours of hard labor in the fields or factories, as did our ancestors. They benefited from their physical fitness, but may not have benefited from their diets. For most of us today the opposite is true. We have the benefit of better diets for preventing poor health, but we neglect the benefits of adequate exercise as a means of preventing illness and promoting a healthy body.

Prehistoric people were too busy surviving to be much concerned with physical fitness, but no doubt they were fit because only the "fit" survived. Once agriculture and the domestication of animals became common, and cities developed, people had more time to do things other than just hunt for and gather food. One of these things was to engage in games that required physical skills and a healthy body.

The ancient Greeks developed organized sports into the Olympic games, held every 4 years in Olympia, as far back as the eighth century B.C. In the year 393 the Olympic games were banned by the ruling Christian Emperor, Theodosius I, because he considered them a pagan ritual. The Olympics were revived in 1896 and held in Athens, Greece.

Physical fitness is defined as the ability of the human body to meet demands imposed on it by the environment. This ability is further defined as that level of fitness required to enable a person to carry out daily tasks with vigor and alertness, without undue fatigue. Today physical fitness is usually determined by tests that measure strength, endurance, agility, coordination, flexibility, and the body's reactions under physical stress.

The types and styles of exercise required to maintain physical fitness are somewhat determined by one's age and lifestyle. For instance, a young person may engage in a vigorous exercise program that includes running, cycling, boxing, wrestling, vigorous aerobics, and team sports. For an older person, exercises such as running or vigorous aerobics may not be appropriate because of the strain caused to joints and the cardiovascular system. As one ages, walking, swimming, cycling, dancing, golf, bowling, and so on may be more appropriate.

Over the past 25 years aerobic exercises have become extremely popular. The main purpose of aerobics is to help men and women increase cardiovascular fitness by improving their bodies' use of oxygen without placing undue strain or stress on their hearts. Most aerobic exercise programs involve about 20 minutes of vigorous movement that should be repeated at least three or four times weekly. It is possible to begin your own physical fitness program with what is called a "low-impact" aerobics which reduce the chance of injury to ankle, knee, and hip joints.

The types of exercise promoted by physical fitness experts change with the times. As a result many misconceptions prevail. One misconception is that a strenuous exercise is only beneficial if muscles stain and ache, encouraging one to endure discomfort. Pushing oneself may be appropriate training for an athlete, but it makes no sense for the average person trying to increase muscle tone and keep physically fit. Another misconception is that to lose weight one can exercise a few times a week while eating his or her regular diet. One of the best ways for most persons to lose weight is by pushing away from the table. Another misconception is that a person should begin an exercise program with the most difficult exercises and work on these until they are mastered. Most experts recommend an initial stress test and an exercise program designed for the individual based on a number of factors, such as age, weight, gender, current physical condition, illness and so on.

ALTERNATIVE MEDICINE AND MODERN MEDICAL MISCONCEPTIONS

Over the ages many alternatives to current medical practices have been proposed. By proving their worth through experimentation some of these practices have been adopted, while other alternative medical practices have not stood up well under the scrutiny of experimental science. Medical beliefs and myths, on the other hand, may or may not be related to any current alternative or borderline form of medicine. Many of these modern medical misconceptions are updated versions of myths dating back thousands of years.

In 1992 the U.S. Congress initiated a mandate for the National Institutes of Health (NIH) to establish the Office of Alternative Medicine (OAM). OAM's main function is to provide clinical research awards to investigate forms of alternative medicine to ". . . facilitate the evaluation of alternative medical treatment modalities" (OAM, 1995, p. 1) for the purpose of determining their effectiveness, and to be a clearinghouse for information through publications and conferences. The following are some of the types of programs being investigated.

Diet, nutrition, lifestyle changes: Follows changes in lifestyle, diet, nutritional supplements, Gerson therapies, macrobiotics, and megavitamin therapy. (An example of funded research: macrobiotic diet as a cure for cancer.)

Mind/body control: Explains art therapy, relaxation techniques, biofeedback, counseling and prayer therapies, dance therapy, guided imagery, humor therapy, hypnotherapy, psychotherapy, sound/music therapy, support groups, yoga, and meditation. (An example of funded research: hypnosis to accelerate healing of bone fractures.)

Alternative systems of medical practice: Investigates acupuncture, anthroposophically extended medicine, Ayurveda, community-based health care practices, environmental medicine, homeopathic medicine, Latin American and Native American health/medical practices, natural products, naturopathic medicine, past-life therapy, shamanism, Tibetan medicine, and traditional Oriental medicine. (Examples of funded research: acupuncture for depression and osteoarthritis.)

Manual healing: Considers acupressure, Alexander technique, aromatherapy, biofield therapeutics, chiropractic medicine, Feldenkrais method, massage therapy, osteopathy, reflexology, Rolfing, therapeutic touch, Trager method, and zone therapy. (Examples of funded research: massage therapy to treat HIV [human immunodeficiency virus] and therapeutic touch to treat stress to the immune system.)

Pharmacological and biological treatments: This includes anti-oxidizing agents, cell treatment, chelation therapy, metabolic therapy, and oxidizing agents such as ozone and hydrogen peroxide. (An example of funded research: pancreatic enzyme therapy to treat cancer.)

Bioelectromagnetic applications: This involves blue-light treatment and artificial lighting, electroacupuncture, electromagnetic fields, electrostimulation and neuromagnetic stimulation devices, and magnetoresonance spectroscopy. (An example of funded research: electrochemical direct current to treat cancer.)

Herbal medicine: This includes ingestion of echinacea (purple coneflower), ginkgo biloba extract, ginger rhizome, ginseng root, wild chrysanthemum flower, witch hazel, and yellowdock. (Examples of funded research: Chinese herbs to treat hot flashes and common warts.)

Let us consider some examples of misconceptions relating to alternative forms of medicine as they developed in the Western world. We will then briefly examine some modern medical myths and misconceptions.

Homeopathy

In 1810 Samuel Christian Hahnemann (1755–1843), a German doctor, published a book entitled *The Organon.* In this book he advocated a form of medicine called *homeopathy,* which states that "likes cure likes." In other words, a drug will cure a specific disease if that same drug, taken by a healthy person, causes symptoms similar to that specific disease. Hahnemann was convinced that the smaller the dose of a drug, the more potent it would be, and thus more effective in curing a specific disease. His major misconception was the belief that as a drug was diluted to greater and greater degree (even to the molecular level), the drug became less *material* and more *spiritual,* and therefore more curative. His method was to continue diluting the toxic substance used as a drug, while vigorously shaking the mixture between each dilution. This shaking he called *succussion.* He claimed that the more his medicines were diluted, the more effective they were, even to the point that there was nothing left but the water and/or alcohol. He called this the *law of infinitesimals.* Hahnemann claimed that the water molecules (solvent) "remembered" the medicine's molecules (solute), even though there was no medicine in the end product (solution). Homeopathic doctors still claim that the effects of the original medicine's molecules are imprinted on the water molecules; therefore, it does not matter that no molecules of the original substance are left in the single water drop that is placed on a sugar tablet and used as medication. Several homeopathic researchers continued this misconception, but went even further. They claimed that the imprinting of the original substance on the water's molecular structure was a "radiant" effect and called it "biophronton" radiation (no such radiation has ever been verified). One researcher even claimed that this electromagnetic link could be used to "potentize" your water medication over the telephone—you did not even need to visit the doctor.

In *The Organon,* Hahnemann states that the fundamental cause of most diseases is "psora," which is know as the "itch" to most of us. His current followers, including the NIH, seem to ignore this part of his misconceptions. One reason for the spread of homeopathy's misconceptions is that several famous people including Oliver Wendell Holmes, William Cullen Bryant, Marlene Dietrich, Ralph Waldo Emerson supported these unproven medical practices. Until recently, the family physician for the British Royal Family was a homeopathic doctor. Since 1900 homeopathy has lost much of its support. Hospitals that at one time treated patients with homeopathic medicine, are now staffed with certified medical doctors who practice proven medical procedures. Although still practiced, homeopathy is not considered a mainstream branch of modern medicine. Even so, the U.S. Government provides many thousands of dollars to support research related to homeopathy and its medicines. The Food and Drug Administration does not regulate homeopathic medicines because of a 1938 grandfather clause that exempts older medicines. A statue of Hahnemann was built on Scott Circle in Washington, D.C.

Naturopathy

Naturopathy exists over much of the world, including the United States. It has no single founder but many believers who insist on relying on nature to heal them. They may use limited surgery or medicines, but most concoctions and therapies are nature related, such as sleeping on bare ground, taking steam baths, walking barefoot on wet grass, sunbathing, and taking no salt or water with meals. Many naturopathists are vegetarians. An early supporter of naturopathy was John H. Kellogg, a Seventh-Day Adventist who wrote a 1,200 page book entitled *Rational Hydrotherapy.* W. K. Kellogg, the "corn flake king," was the younger brother of John Kellogg. Another believer was Henry Lindlahr, who claimed that disease was not caused by bacteria or germs, but rather that the body produced microbes as nature's way of healing itself. Another follower was Benedict Lust, who established a naturopath school and resorts in the United States. Lust was arrested many times for placing false advertisements in nature magazines. His treatments consisted of water therapy, enemas to rid the body of poisons, fasting followed by exercise, and diets of grapes and goat's milk. Naturopathy claimed to be able to cure any disease, including cancer, smallpox, diphtheria, and whooping cough. By insisting on using unpasteurized milk, rejecting sulfa drugs and penicillin, naturopathy caused needless deaths. Naturopathy is not considered a legitimate medical specialty. Even so, research into naturopathic cures is supported by the U.S. Government.

Iridiagnosis and Zone Therapy

Two offshoots of naturopathy and homeopathy are *iridiagnosis* and *zone therapy*. Iridiagnosis is the diagnosis of illness based on the appearance of the iris of the eye. The theory is that the iris is divided into forty zones that correspond to organs and systems of the body. If spots or "lesions" appear in a particular area of the iris, something is wrong with the corresponding body part. Zone therapy assumes that the body is divided vertically into ten zones—five on each side. Each zone ends in a finger or toe. There is no explanation of how or if there is any connection between the zones or body parts—no mention is made of blood vessels, nerves, muscles, and so forth. Followers of zone therapy believe one can cure a particular illness by tying a rubber band around the appropriate toe or finger. Needless to say, zone therapy is not considered mainstream medicine. The U.S. Government supports research into the effectiveness of zone therapy.

Osteopathy

Osteopathy was founded by Andrew Taylor Still (1828–1917), who had no medical or scientific training. Much of his writing is a disjointed hodge-podge of false autobiographical and pseudomedical information that Still attributes to divine inspiration. The word *osteopathy*, as coined by Still, means "sick bones," which basically relates to the small bones in the spine. Because these bones become dislocated they put pressure on blood vessels and nerves preventing the body from curing itself—thus the osteopath must adjust these bones, which will stay adjusted and cure the patient. Osteopathic schools and hospitals have been formed in the United States. Osteopaths claim that their spinal manipulations will cure yellow fever, diphtheria, diabetes, malaria, hemorrhoids, constipation, obesity, and so on. There is no experimental evidence that traditional osteopathy is effective; many of the small osteopathic hospitals and their physicians have accepted mainstream medical concepts and practices. The NIH supports research that investigates some of the more up-to-date osteopathic concepts.

Chiropractic Methods

Chiropractic methods and procedures comprise another type of alternative medical practice. Chiropractic was founded by Daniel David Palmer (1845–1913), who in 1885 believed he could cure people with his "animal magnetism," which involved spinal manipulations. There are over 30,000 practitioners of chiropractic manipulations worldwide; these chiropractors have a rather loyal following. Unlike osteopathy, which changed over time, chiropractic practice has not. Some chiropractors claim to cure just about everything, including deafness, scarlet fever, diphtheria, and so on by manipulating specific vertebra in the spine and neck. Most devotees, including some professional people and physicians, seek the services of a chiropractor when they have back problems. Spinal manipulation, primarily for back problems, is how most chiropractic businesses advertise and market their services. In the United States the National Chiropractic Association, with over 20,000 members, is the practice's main organization. The U.S. Government supports research to investigate the effectiveness of chiropractic practices.

Some Individuals and Their Theories

Another way to look at these misconceptions concerning alternative medicine is to consider some of the individuals who were responsible for them. At times they are referred to as medical charlatans, quacks, or just plain frauds.

Let us start back in 1796 when Dr. Elisha Perkins patented a device consisting of two, 3-inch rods, one of a copper, gold, and zinc alloy, the other of silver, platinum, and iron alloy. This device was called "Perkins' Patented Metallic Tractor," which when drawn over the ailing body part, would just "draw" off the disease. He sold his "tractors" to George Washington and Chief Justice Oliver Ellsworth. Perkins' son Benjamin made a fortune selling his father's invention.

In 1910 Dr. Albert Abrams, who had an orthodox medical career, published several books that led him down a mythical path. He developed a technique called "tapping." His first technique involved only the spine, but he finally perfected a method of tapping the abdomen. His theory was that every disease had its own "vibrating rate" that could be diagnosed through tapping. In 1920 he invented an "oscilloclast," which made use of these vibrations. He even invented the "reflexophone," which could be used to diagnose diseases over the telephone. The American Medical Association sent a blood sample from a guinea pig (named Miss Bell) to Dr. Abrams for analysis. The diagnosis came back that Miss Bell had a sinus infection and a streptococcic infection of her left fallopian tube. In another test a Michigan doctor sent the blood of a rooster to Dr. Abrams. The diagnosis indicated that the chicken had malaria, cancer, diabetes, and two venereal diseases.

Since Abrams's time, many medical charlatans have built better "mouse-traps." One such device diagnosed illnesses using the "vibrations" of a small blood sample inserted into a machine similar to the one invented by Dr. Abrams. It could be used over the telephone and, needless to say, it never passed an objective test by scientists.

Radiation became a cure-all for many myth makers. One type was called *vrilium* and was invented by Abbott E. Kay. Its name came from the word "vril," which he said came from cosmic energy that was used by a superior race in space. One could buy a small metal vial of vrilium from Robert T. Nelson. If pinned to one's clothes, it would kill all germs, relieve pain, and cure diseases. Vrilium cured Nelson's poverty problem.

Electric devices, including radios, ozone emitters, and electric/magnetic "stimulators," were designed for many purposes and are still favored by medical frauds. One such device, invented by Dr. Fred Urbuteit, shoots a low voltage electric current though the ailing part of the body. It cost its users $1,500 to $3,000 each. Another famous "cure" used colored lights. In 1920 Colonel Ghadiali of New Jersey discovered "Spectro Chrome Therapy." When used with special diets these colored lights could cure specific diseases by "tonation." One had to use it with one's head pointed north, and the use of the correct color of light was essential if the condition was to be cured. Diabetics were to eat lots of starch and sugar and interchange yellow and magenta lights. A blue-green light is required to cure gonorrhea. Bright red light increases sexual desire, whereas purple reduces sexual desire. Colored-light therapy has a long history of varied practitioners and many followers who made the good "doctors" rich.

Cancer

Cancer seems to attract a great variety of unproven mythical medical practices. The *Complete Home Medical Guide* (1985), written by a group of physicians from the Columbia University College of Physicians and Surgeons, states:

It is almost unthinkable that unscrupulous individuals would set out to profit by offering desperate cancer patients treatments that are of little or no value. The sad reality is that cancer quackery is a multibillion-dollar-a-year business and many thousands are misled into turning to useless unproved treatments. . . . The laetrile clinics in Mexico, unproved treatments with megavitamins, gadgets that have been rejected by federal or other regulatory agencies or the medical establishment are but a few examples of the kinds of cancer quackery that continue to thrive. (p. 402)

The book lists some questions that one should ask before entering a questionable treatment: *"Is there a verifiable track record? Has the treatment been reported in the medical literature or reputable publications?"* The text continues: *"Be wary of self-published booklets . . . and reports limited to sensational popular publications or publicity statements"* (pp. 402–3). One is also advised to check the practitioner's background and the clinic or hospital in question with your local medical association.

One of the earliest cancer cures was proposed by Dr. William F. Koch, who in 1919 "discovered" a cure-all called *gloxylide*. He stated that it was a

Chapter 3

Life Sciences

INTRODUCTION

Scientific developments and misconceptions about life include the most compelling questions asked by people of all cultures, throughout the ages, including the present. Most misconceptions about life arise from superstitions, myths, ignorance, beliefs, dogmas, theories, or scientific laws that relate to the meaning and origin of life. Numerous explanations have been proposed to support equally numerous "beliefs" and "misconceptions" about the nature of life. In this chapter we will explore some of these ancient and modern beliefs and misconceptions. But first, a bit of background on what many scientists currently believe about the origin of the universe, our solar system, the Earth, and life on Earth.

One of the accepted theories for the beginning of the universe is called the *Big Bang* in which an incredibly dense, small ball of mass exploded and expanded rapidly outward, producing a higher temperature than our sun's. The explosion was forceful enough to overcome the gravity of the particles that expanded, and that are still expanding after 15 billion years. The particles and energy are what make up everything in the universe (see the Big Bang Theory in this chapter). Where the original "matter" for the Big Bang came from, or exactly of what it consisted has still not been determined. Besides which, we do not know if this is really what happened. It is a theory that was worked out "backwards."

Figure 3.1 provides a time line of progressive occurrences. As you can see, life on Earth must have been somewhat unique in time as well as space. But with the vastness of the universe, there must be hundreds, if not thousands, of solar systems that under the proper conditions could develop life. Some

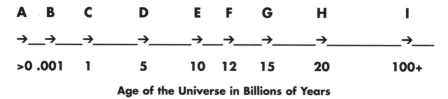

Figure 3.1 Age of the Universe in Billions of Years. All figures are estimates in billions of years. Not drawn to scale. **A.** Beginning of time and the universe after the Big Bang. **B.** One hundred million years after the Big Bang, when it became cool enough to form atoms from the protean particles (mostly hydrogen nuclei). **C.** The period when galaxies were formed, about 1 billion years after the Big Bang. **D.** Five billion years after the Big Bang, when the stars were formed in the galaxies. **E.** Our solar system, including the Earth, was formed at about 10 billion years after the Big Bang. **F.** At about 12 billion year after the Big Bang the first simple *micro*scopic life forms appeared, followed by single-celled plants and animals. **G.** The development of *macro*scopic life, including modern plants and animals. This represents the present, which is about 15+ billion years after the beginning of the universe. **H.** At about 20+ billion years our sun will greatly increase in size as it cools, turn into a giant red star, and engulf the entire solar system, including the Earth. **I.** At 100 billion years all stars will cool and die, bringing on the threat of death of the universe. Beyond 100 billion years, possibly around 2000 billion years after the Big Bang, the Big Crunch occurs and the entire universe collapses into itself—this assumes that the universe is a closed system. But if the Universe is an open system it will continue to expand and regenerate itself forever.

consider it arrogant for humans to assume that our Earth (or solar system) is the only possible place for life in the universe. Let us examine some of the conditions required for life as we know it to develop. And what the differences are between living and nonliving matter.

The following are some of the conditions that are required for a planet to develop and sustain Earth-type carbon-based life forms.

1. The planet must not be too close, like Mercury; or too far away, like Pluto (or even Venus), from its central sun. The temperature range for life on Earth is very narrow.

2. Energy is involved when smaller atoms form larger molecules of compounds. Our sun's energy is the main source of most of the energy on Earth, but radioactivity, volcanoes, and lightening are also sources of energy and heat on Earth. According to the laws of thermodynamics, all organized systems tend to either reach energy (heat) equilibrium or they randomly lose energy, run down, and become more disorganized, with increasing chaos (i.e., entropy)—unless additional energy is available to reverse the process so as to form more organized systems such as life. Without the input of energy, systems, including living organ-

isms and molecular compounds, will proceed from the more complex to the simple, and from the more organized to the more disorganized, and become random or chaotic. In other words, disorganization results when energy is not provided to maintain the system; death is an example of this. The thesis indicates that with the input of energy simple molecules combined in a fashion to produce larger molecules, which, in time, developed an autocatalytic ability to reproduce themselves. The ability to reproduce is *the* main criterion for all life. A depressing thought is that, as far as nature is concerned, once an organism has reproduced adequate offspring to continue the species, nature doesn't care at all about the aging parents.

3. The size of the planet is important. If it is too large, its gravity will prevent certain light gases from escaping that could inhibit the formation of life, as with the example of methane on Jupiter. But if the planet is too small, like Mars, most of the gases essential for life—hydrogen (H_2), oxygen (O_2), nitrogen (N_2), and carbon dioxide (CO_2)—will escape its gravity (except for possible traces).

4. Oxygen was produced from trillions of microscopic organisms and simple plants. Oxygen and several other gases, such as chlorine (Cl_2), are acted on by the sun's energy in the upper atmosphere producing ozone (O_3), which blocks harmful ultraviolet rays from reaching the Earth's surface. Excessive ultraviolet energy (UV) would be expected to prevent the evolution of higher forms of life, but UV rays as a form of electromagnetic energy may also have been responsible for the beginning of life, mutations, and therefore evolution. The book *Slanted Truths* (Margulis and Sagan, 1997) mentions that of the three similar planets in close proximity, that is, Mars, Earth, and Venus, only Earth has developed an oxygen-rich atmosphere. Theoretically, Earth should have a high level of carbon dioxide as does its two close neighbors: it probably did in its early evolution, which promoted plant growth.

5. There are other suggestions for the origin of life on Earth. Both Lord Kelvin (1824–1907) and Svante Arrhenius (1859–1927) postulated that life on Earth was "seeded" (called *panspermia*) from either meteorites, extraterrestrial spores, or other microorganisms via cosmic dust.

6. There are many theological, philosophical, and mythological explanations for the origin of life. Unfortunately, none have any empirical evidence to back them up.

There is a narrow band of crucial conditions that can support the carbon-based life found on the Earth. Speculations about a possible silicon-based form of life, instead of our carbon-based form, have been made in the past, but this has not been proven. One possible problem is that silicon, although chemically related to carbon, forms crystals rather than the liquid type of complicated rings and chains of organic molecules required for life as we know it.

What is the difference between something that is living (animate) and something nonliving (inanimate)? Another way to ask the question is: What do living things do that nonliving things cannot do?

First, we will compare the chemical anatomy of rocks and living matter. Both are composed of certain types of atoms and molecules. Rocks are mainly composed of silicate minerals, which are the major building blocks of the Earth's crust. Of importance is that the atomic and molecular composition of rocks, once formed, have a relatively fixed chemical composition. The anatomy of living things is also composed of atoms and molecules, but the chemical composition is not permanently fixed. An adult human is composed of over a thousand million, million, million, million atoms. There are about 25 different chemical elements found in the human body. Only oxygen, carbon, hydrogen, nitrogen, calcium, and phosphorous are found in quantities larger than 1%. It is estimated that the going market price for all the elements in one's body is $9.95.

Ancient people often gave some "living" characteristics to inanimate objects. Many cultures also thought that once a living thing or person died it still held some form of life that would return. Thus arose the practice of mummifying the dead and building elaborate burial vaults containing items needed for the next life.

All living things have what is called a *life cycle,* which extends from birth to death. The two most important characteristics of this cycle are *reproduction* and *metabolism.* No nonliving thing possesses these characteristics.

Reproduction is divided into two types: *asexual* and *sexual. Asexual* reproduction occurs when a new generation is produced by division of the cells of one parent, by budding from the parent, or the formation of spores by a single parent. Asexual reproduction does not require the cell nuclei from two different sexes to combine, thus each new individual formed by asexual reproduction is identical to its parent. *Sexual* reproduction requires the union of the nuclear material of a male sperm cell with the nucleus of a cell from a female egg. Thus, the new individual plant or animal has characteristics of both parents.

Metabolism is the sum of all the chemical and physical reactions found in living organisms, their biological systems, their organs, their tissues, and cells. There are several crucial anatomical structures and physiological processes exhibited by all living things. They are as follows:

1. *Reproduction,* which is required for the continuation of the species.
2. *Respiration*—in animals, the addition of oxygen to the blood for the production of energy and the removal of CO_2 and excess water vapor; in plants respiration occurs as opposite gas exchange plus transpiration (release of water vapor through the leaves).
3. *Digestion,* which includes ingestion of food (eating), the chemical breakdown of food and assimilation of the digested food. Plants convert "food" into carbohydrates.

Ancient Babylonians

At about 1,200 or 1,300 B.C. during the reign of the early Babylonian king, Nebuchadnezzar, creation myths were attributed to their high god, Marduk. As with Egyptian, and later Christian stories of creation, the Babylonians included the formation of the universe out of chaos, the creation of the Earth and its matter from a primordial "mother," who placed water on Earth to bring forth life. Although the high god created other gods, in time there became only one god who was both the creator and destroyer. This led to the concept of one creator or monotheism, which brings us to Genesis.

In about the fifth or sixth centuries B.C., the Babylonian concepts were adapted and adopted by the Hebrews as their cultures mixed during various wars. The Hebrew story of creation involved the chaos of water and "in the beginning God created the heaven and the Earth," et cetera. From here the well-known story of Genesis continues. It is somewhat unique because this was the first time that a human, not a mythological being, was directly involved as part the creation myth.

Ancient Indians

The concepts of creation from the myths of India date to about 2,000 B.C. Like Egypt, the Indians had several gods and similar creation myths. One of their main gods, Brahma, caused the creation of everything by again spilling his seed to form the cosmic egg. From this egg, their god created everything out of nothing, including life-giving water, the Earth, the sun, the heavens, night and day, and everything else. One difference was the concept of the Indian "mind," which was considered the human soul.

Ancient Greeks

Greek creation mythology dates to about the eighth century B.C. and is recorded in Hesiod's books. Homer included the classical Greek creation myths in his writings. The Greek myths included many from other civilizations, which is to be expected as cultures mixed when wars were fought. The common thread was the creation of the universe and everything in it out of chaos. The ancient Greeks established a single "god" in heaven, but their mythology included many lesser gods and goddesses, including animals. Females were recognized as equals, as demonstrated by the myth of Mother Earth (Gaia) giving birth to life. As Greek civilization diminished, Roman culture flourished, and much of Greek mythology was converted into Roman myths and gods. For example, the Greek Earth goddess, Gaia, became Terra in Roman mythology. The Greek goddess of beauty and love, Aphrodite, became the Roman love goddess, Venus.

Christians

Many Christian beliefs are the outgrowth of these older mythologies. The early Jewish history as reported in Genesis, the first book of the Old Testament, starts with phrases similar to those in Egyptian myths, "In the beginning God created the heaven and the Earth" (Genesis 1:1). And, "He moved upon the face of the water" (Genesis 1:2) The Christian God created everything in steps, and by his "word," not just something out of nothing as with chaos. There is some connection between the myth of the High God of Egypt and the belief of Jesus being divine, and the idea that all things are God's, including the beneficence of the Earth and retribution.

Native Americans

A lesser-known creation myth is that of the Hopi Spider Woman. The Hopi myth is associated with many other creation myths with a few important distinctions. First, the creative forces in the Hopi mythology were female. Tawa was the Sun God, and Spider Woman was the Earth God. Even today this matrilineal arrangement is part of various American Indian cultures and is expressed in tribal dances, rituals, and crafts. Second, is the devotion to nature and the environment that is inherent in the mythical and spiritual associations of Native American tribes. See Chapter 6 for more on the environmental myths of ancient American Indians.

Bantu

The Bantu myths of creation may be much younger than the others mentioned. Their equivalent to the Egyptian High God is Bumba, who vomited to create everything from nothing, including the sun, the moon, the stars, and then the Earth and living things. Oddly, Bumba was not only a male god but also white.

Common Elements of Creation Myths

Several elements common to creation myths are as follows: (1) Presence of the highly respected "parents" of the Earth, or the Earth Mother or Father; (2) the idea of the "cosmic egg"; (3) the "seed" from which everything was formed; (4) the concept of water (pregnant chaos) and the flood, which created life anew; (5) darkness yielding to light; (6) the movement from a void or nothingness to developed order; and (7) the dualism of greater to lesser spirits, as well as monotheism—the concept of a single supreme being, often represented by a revered human.

SCIENTIFIC CONCEPTS OF CREATION

We now turn to modern scientific creation stories. Like the myths of other cultures we have our own myths, which we like to think are based on more rational premises arrived at through universal physical laws, critical observations, measurements, and predictions.

The Big Bang Theory

The current cosmological creation story, accepted by most scientists, is called the *Big Bang Theory*. As previously discussed, this theory posits that at about 15 to 17 billion years ago there was a huge explosion of a very, very dense, small ball of matter. This ball contained all the energy, all the elements, and all the gravity that now exist in the universe. This explosion was so violent, producing such great heat and energy, that all the particles of matter escaped from this dense ball's gravity and started spreading outward in all directions. As time progressed, this outward spread did not slow down but rather accelerated, as it still does today. About 5 billion years after the explosion, many of the particles, which were mostly hydrogen nuclei (protons), formed the stars. As things cooled, atoms of various elements (mostly helium with two protons in its nucleus) were formed. The stars, not the Big Bang, formed the more complicated elements with higher atomic weights. Ten billion years after the Big Bang our planetary system was formed by gravity "collecting" material from space, also forming our sun and galaxy in the process.

About 3.5 to 4 billion years ago atoms and molecules combined in groupings or networks. This type of combination is a normal atomical process in which the combined atoms reach a lower state of entropy but do not progress to a state of disorganization or complete randomness (entropy). This is the physics of what drives atoms to combine according to their valences, which we know as *chemical reactions*. These networks developed in the primordial or prebiotic soup of the Earth, some became what is called *autocatalytic*, which means they can cause other networks to grow and divide. This is one concept of how life began. From these chemical and physical beginnings, very simple biological systems developed through a process of self-organization, which led to more complex systems. From these very early, very simple forms of life, single-celled organisms developed, followed by multiple-celled organisms. The reactions of atoms and molecules continued within these cells, some of which developed nuclei, which are also made up of chemicals obeying physical laws.

The first appearance of very simple microscopic life formed on Earth about 2.5 or 3.5 billion years ago, but it was not until about 500 million years ago that multicelled land plants evolved from this primitive life. Flowering plants did not evolve until about 125 million years ago. Amphibians and reptiles evolved about 500 million years ago. Mammals about 60 or 100 million years ago. Prehistoric man (*Homo erectus*), who walked upright on two feet, developed simple tools, and used fire, evolved about 500,000 years ago. Modern man (*Homo sapiens*), the wise man, developed no more than 20,000–35,000 years ago, which is a very short period of time compared to the age of the Earth and universe.

Theories of the Ancient Greeks

Several specific beliefs and misconceptions about creation have been recorded. In this section we will consider some of these.

The Greek philosopher Thales of Miletus (c.625–c.547 B.C.) was the first to break away from supernatural explanations by trying to understand the physical makeup of the Earth and everything on it. His big misconception was that everything on Earth floats and that everything is made from or composed of different forms of water. This included the mountains, air, sky, plants, animals, and people.

Anaximander (c.611–c.547 B.C.), also from the Miletus school, disagreed with Thales of Miletus and doubted that the world and all its contents were made of water. His big misconception was that he believed the Earth was unsupported at the center of the universe—that it just "hung" there. He correctly guessed that all life originated from the sea, that the Earth's surface was curved, but he thought it was like a cylinder, and that the heavens were a complete sphere encircling the Earth, not just a semisphere. He was the first to map the "world" of Greece.

Anaximenes (c.570–c.500 B.C.), also a follower of Thales of Miletus, held the misconception that the Earth was created from a formless substance he called "aperion," which he considered to be the source of all things. He stated that air was the fundamental element and when air is compressed it forms water and the Earth. When heated, air became fire.

Plato (c.428–c.347 B.C.) stated that it was possible for humans to acquire knowledge. One of his stories that might be considered a creation myth is one in which he describes a cave where shadows cast by the sun are all that can be seen on the walls of the cave (all one can know about the world) until they break into the *daylight* (a metaphor for *knowledge*). He was concerned with "forms" and their hierarchical arrangements, which led to misconceptions about living things and the physical environment.

Aristotle, along with Plato and other Greek philosophers, has had the greatest influence on Western thinking of any individual. Aristotle's scientific misconceptions include his belief that the world was "born" of and composed of just four elements: earth, water, air, and fire. Nevertheless, Aristotle was the first to use observation and deduction so as to determine why things happen. Even though his ideas contained many errors, he was the first to provide an overall, integrated system of how life, the world, and the universe were created and how they work.

There are many other ancient Greek thinkers who made speculations as to the nature of creation and the universe. We will name just a few. Pythagoras (c.580–c.500 B.C.) believed that a fire existed on the other side of the Earth that we could not see but that was reflected by the sun, and that life was renewed each day, just as the sun reappears each day. Heraclitus (c.540–475? B.C.) believed in opposites and that everything was in turmoil. He said that all things undergo change and that fire, the major element, caused these changes, which included new life, similar to the sun being replaced every day. He is the one who made the classical philosophical statement, "A man cannot step into the same river twice." He also said, "There is nothing permanent except change."

Spontaneous Generation

As late as the seventeenth century, people assumed that life could just "start" in things like decaying food, urine, and manure because worms or maggots could be seen hatching there after a few days. They had the misconception that rats were spontaneously generated in piles of garbage, and that animals that lived in slime and mud were generated by it; for example, frogs, salamanders, clams, and crabs. One of the first scientifically controlled experiments to test this "theory" was conducted by Francesco Redi (1626–1697). He placed meat in eight jars, four unsealed and four sealed. The meat that was not exposed to air did not generate maggots, but the exposed meat did develop maggots. He thought that the air may have something to do with the results, so he next placed gauze over several jars containing meat and left several other jars filled with meat open to the air. No maggots appeared in the jars of rotten meat that were covered by the cloth, which could let air in, but not flies. Maggots did appear in the open, uncovered jars of rotten meat even though the fly eggs were too small to see.

Unfortunately, people who were told or read about these experiments did not believe the results, so if they wanted to believe in spontaneous generation, they still did. The concept of *spontaneous generation*—life "springing from nothing"—was finally disproved to most people's satisfaction by Louis

Pasteur (1822–1895). Pasteur demonstrated that air carried dust and micro-organisms, which if sealed out of containers of food, or killed by heat, would not spontaneously generate life.

In 1953 Stanley Lloyd Miller (1930–), a student of the Nobel laureate chemist Harold Urey (1893–1981), tried to "create" life from some elements and compounds thought to be the precursors to life. He mixed hydrogen, ammonia, and methane gas in a chamber with pure water and then passed an electric charge through the gases to add energy similar to lightning. After several weeks he found several organic compounds and simple amino acids. This was the first time any amino acids of life were formed outside a living organism. Other scientists experimented with different combinations of gases and "soups" of chemicals, which resulted in the "creation" of other organic compounds. To this day no one has formed life in the laboratory, not even molecules of DNA (deoxyribonucleic acid), which are required for living things to reproduce—making DNA a criterion for life. But a start has been made, and based on the evidence, it may be possible for life to develop and evolve without a supernatural cause.

Other Scientific Theories

There are two rather new beliefs, or possible misconceptions, as to the origin of life. These theories have not yet been proven. One of these involves organic material embedded in eight meteorites that originated on Mars and was found on Earth. This discovery has led people to speculate that the source of life on Earth could have come from Mars. Recently, dozens of snowball-like comet chunks of icy materials have been detected entering the Earth's upper atmosphere every minute. They vaporize and add water to the upper atmosphere. It is possible, if they may also contain organic material, that similar ice chunks could have been the origin of life.

A rather new, controversial theory called the *Gaia hypothesis* is advocated by Lynn Margulis and Dorion Sagan (1997), among others. The Gaia hypothesis, although based on factual information about the Earth and life, also has some metaphysical aspects to it. The Gaia theory includes the atmosphere and surface sediments of Earth as a whole. Gaia is the region of life on Earth (also thought of as Mother Earth, the biosphere, or the area of the biota). One aspect of the Gaia concept is that the Earth is "living." Several components of the Gaia hypotheses are as follows:

1. The Earth's temperature, composition, and atmosphere form the regulated biota (all living things on Earth).

2. The biota has been in existence since the beginning of life on Earth (about 3 to 4 billions of years ago).

3. The continuation of biota on Earth is not a matter of chance.

4. The atmosphere is an extension of the biosphere—it becomes its own environment.

5. Life interacts with and controls physical aspects of the Earth on a global scale.

6. The composition of the reactive gases—for example, oxygen, nitrogen, hydrogen, and their temperatures—have remained somewhat constant for many billions of years.

7. Gaia makes its own environment, regardless of the many changes over the years.

8. Gaia will remain inhabited by many forms of life (biota) for eons to come through evolution and by adapting to changing conditions.

CHANGE

Most ancient people thought things in the world stayed more or less the same. They could see that the stars were the same, that the moon and planets moved with regularity, that the different seasons returned each year, and so forth. Heraclitus was the first person to propose that *all* things change, and nothing is permanent. He was known as the "weeping philosopher" because of his belief in constant change—no one could count on anything ever being the same. He stated, "A man cannot step in the same river twice." By this he meant that the river is always changing—the water (molecules) is not the same water first stepped into, and the water causes erosion to change the river bed, banks, and gravel. Possibly more important is the thought that stepping into the river the first time is a new experience for a person, and therefore the second time the person steps into the "same" river the experience cannot be the same because the person has "learned" from the original experience and is no longer the same person. We know one thing for sure, and that is *the only consistent thing in the universe is change.*

Because people did not see the Earth changing in significant ways, such as mountains being formed, they had a difficult time conceiving of the evolution of inanimate objects over time. They knew that plants and animals were "born" from other plants and animals, and that the offspring was similar to the parents.

Scientifically there is no question that both living and nonliving things change or evolve. Many nonscientists doubt that living organisms have evolved or are still changing. To consider the process of evolution it is necessary to examine the physical and chemical requirements for life. So let's take another look at the formation of the Earth and the conditions that made life possible.

As previously mentioned, the Earth was most likely formed about 4.5 to 5 billion years ago, which was about 10 billion years after the Big Bang or cosmic egg theory, estimated at about 15 billion years ago. No one has a scientific explanation as to where the original tiny, dense ball about the size of BB shot came from. Even so, according to modern theory this "ball" created all matter and energy that exists in the universe. Most cosmologists consider that the tiny ball "just was" or they may provide some metaphysical explanation. There are other scientists who consider the universe in a state of continuous generation of new matter and energy that are forever expanding. In other words, new galaxies and stars are always being created—therefore there is no beginning and there is no end to the process of creation.

The Earth's size, and therefore gravity, was just adequate to hold onto some gases and permit lighter ones to escape into space. From the time of the ancient Greeks until the Middle Ages it was thought that the sun or water provided all the energy, including heat energy, required for life and geologic change on Earth. In 1752 Jean-Etienne Guettard (1715–1786) came up with the theory that it was heat inside the Earth that caused geological changes and thus made conditions on Earth possible for life. This was considered the *Pluto theory*—it was named after the Greek god of the underworld. Most other scientists of this period held the misconception related to the Neptunist theory, which stated that the cause of geological changes and ultimately life, was water, not heat.

In the late 1700s the misconception was held that the surface of the Earth radiates more heat into space at night than it receives from the sun during the day. If this were true, the Earth would have been a frozen planet eons ago. In the early 1900s Clarence Edward Dutton (1841–1912) determined that radioactive elements in the deep layers of the Earth generate tremendous heat, enough to create volcanoes and earthquakes, which keeps the Earth's core molten. Miners working in very deep mines experience temperatures over 120°F. This internal heat is conducted to the surface and then radiated into space along with the heat the Earth receives from the sun. Both the sun's heat and the internally generated radioactive heat help maintain the critical temperature required for life on Earth.

There are several theories of how life started on Earth; some are based on various sciences, others are based on mysticism and theology. Just when life started and how it changes are two of the major conflicts between these theories. For answers, scientists rely on observational evidence found in geology and fossils, whereas some nonscientists rely on philosophy, theology, and ancient writings. The differences might be thought of as *geological* changes over *time.* Let's take a look as some beliefs and misconceptions about geology, fossils, and time.

GEOLOGY

Geology is the study of the origin, history, shape, composition, and age of the Earth. Along with related sciences such as geoscience and earth sciences, scientists use physics, chemistry, biology, and paleontology to better understand what the Earth is and how it relates to life.

Early people surely were curious about the Earth, its mountains, water, rocks, dirt and minerals, its size, and so forth. Geological features such as mountains, canyons, earthquakes, lakes, rivers, deserts, and so on were thought to be the work of gods. They both feared and wondered at their environment, which resulted in a personalization of features of the Earth. Records that indicate a curiosity about geology go back to the mythologies of the Grecian, Roman, Egyptian, Sumerian, and Babylonian empires that existed several thousand years ago.

The Formation of the Earth

Previously several false misconceptions of how the Earth came to be were listed (see "Prehistory and Creation Myths"). They represent several beliefs and misconceptions people held about the formation of the Earth until the seventeenth and eighteenth centuries, when people began to observe nature without the prejudices of myths and biblical history. Following are five misconceptions that were developed to explain how the Earth was formed.

In 1611 Simon Marius (1570–1624) viewed a great hazy patch in the Andromeda constellation through his telescope. He named it the Andromeda nebula (*nebula* means "cloud" in Latin). Later in 1694 the Dutch astronomer and lensmaker, Christiaan Huygens (1629–1695), using an improved telescope, saw another cloudy patch in the constellation Orion, which he named the Orion nebula. A number of scientists of that day speculated that these swirling nebulae were condensed by gravity to form stars. The astronomers postulated that as the nebula became smaller they would speed up their rotation (the law of conservation of angular momentum) and spew out material that formed the planets, including the Earth.

In 1694 Edmund Halley (1656–1742), who discovered Halley's Comet, claimed that the comet's tail caused the biblical flood. In 1696 William Whiston (1667–1752) based his theories on biblical passages that were accepted by the orthodoxy of Protestant England. He accepted Halley's beliefs as well as that the spherical nature of the Earth was formed by the tail of the passing comet. Whiston referred to this process as the original "chaos," which occurred at the time of the creation of the Earth. He believed that the orbits of the planets were perfect circles—not ellipses, that a year consisted of 360 days, that the moon's "month" was 30 days (it actually takes the moon

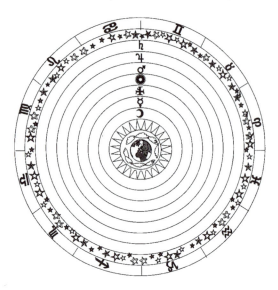

Figure 3.2 An ancient conception of concentric spheres with the Earth in the center, surrounded by multiple spheres containing the planets, stars, and zodiac.

27.23 days to circle the Earth, but it takes 29.53 days to go from one new moon to the next because the Earth is moving at the same time the moon is revolving around the Earth). Whiston believed that a comet started the Earth rotating on its axis. He also believed that the heat of the comet's tail contained evaporated water, and as it cooled and condensed it produced the Flood, which changed the Earth's orbit to an ellipse and increased the length of the year by 5 days. All this water was, he said, drained into the center of the Earth leaving oceans, lakes, and rivers.

Georges-Louis Leclerc de Buffon (1707–1788) suggested that perhaps the Earth was born from the sun. This led to several misconceptions, such as the idea that another big star or comet passed our sun, and gravity broke off chunks of each, which formed the Earth, the moon, and other planets. This belief continued with the idea that as these two giant bodies passed they produced a "swirl" that set the preplanet material in motion, orbiting the sun. Gravity caused this material to consolidate into planets as we know them today. These two theories were the best known and were accepted for many years, and are still argued today.

In 1882 Ignatius Donnelly (1831–1901) described several pseudo-scientific ideas. One was that a comet that had passed the Earth caused catastrophic effects. He drew several erroneous conclusions from this event, which led to his belief in the myth of the existence of the lost continent of Atlantis.

Figure 3.3 The flat-Earth concept with a crystal semisphere containing the heavens that prevented the oceans from "spilling" off the edges.

In 1914 Hans Horbiger (dates unknown) made many mystical anti-intellectual statements on geology, the formation of the Earth, and other "science" he based on his dreams. He also based his work on over 500 ancient myths and legends such as people living in the center of a hollow Earth, the continuous creation of new life forms by a spiritual being, and the passing of a comet that created life and the great biblical flood. Over 1 million of Horbiger's followers developed a cult called "WEL," which is currently located in Germany and England. None of his theories held up to scientific criteria.

The Shape of the Earth

Since men and women began questioning things, they thought the Earth was flat because that was how they observed it. The mountains and waves meant that it was not all smooth, but where it was smooth one could see a great flat expanse in all directions. Eudoxus, Plato, Aristotle, and others designed an Earth that was like a sphere enclosed by other spherical glass globes, which contained all the other heavenly bodies. To account for the movement of these celestial bodies, Aristotle's concept required 27 concentric glasslike spheres surrounding the Earth. Other theories required 50, or even as many as 77 spheres to accommodate models of the solar system and universe (see Figure 3.2). Others, who considered the Earth to be flat, used semispheres as the huge glass celestial domes that extended to the edges of the Earth, where they prevented the oceans from spilling off into nowhere.

This concept of a flat Earth was refined and persists to this day, as evidenced by the over 3,000 members of the Flat Earth Society located in California (see Figure 3.3). The proponents of both a flat Earth and a round Earth use the same passage in the Bible that states, "It is He that sitteth upon the circle of the earth" (Isaiah 40:22), to prove their particular point of view.

The Greek philosopher Pythagoras (c.580–500 B.C.) was the first person on record to state that the Earth was a sphere. He observed how some of the stars and planets moved and the appearance of the curved shadow of the Earth on the surface of the moon during a lunar eclipse. Also, when one climbs a tall mast of a ship during calm weather, it is possible to see a slight dropping of the horizon—particularly when another distant ship seems slowly to disappear from the hull first, not all at once as it would if it just dropped off the edge of the "flat" Earth. Recent photographs of the Earth taken from manned satellites provide vivid proof that the Earth is a sphere, but not a perfect sphere.

We do know that the law of centrifugal force affects all the Earth's particles, but the force is greater at the equator than at the poles because the speed of rotation is greater (about 8,500 mph) at the equator. Thus, there is a slight bulge, making the Earth's diameter slightly greater at the equator, but not enough to worry about. It might be mentioned that a point on the equators of the larger planets, such as Jupiter and Saturn, have a centrifugal force greater than Earth's, producing greater equatorial bulges.

The Size of the Earth

As long as people thought the Earth was flat, there was not much concern about its size. Everyone knew that the maps showed that oceans surrounded the known land. You just kept going and would never fall off because the Earth continued forever. But some were fearful of sailing off the edge. Once a few people realized and accepted the fact that the Earth is a globe, they began wondering just how big it was.

It is recorded that in about 245 B.C. the Greek philosopher Eratosthenes (c.276–194 B.C.) compared two perfectly perpendicular rods, one in Syene (also spelled Cyrene), now known as Aswan located on the Nile river in upper Egypt, and the other in Alexandria (located in northern Egypt on the coast of the Mediterranean Sea). Syene was a good distance south of Alexandria, and he knew that at the summer solstice the sun shone directly into a well in Syene. Thus, he figured that Syene was close to a direct line between the sun and the Earth's center on June 21st. This is about 23.5 degrees north of the equator and is known as the Tropic of Cancer. At exactly noon on June 21st he measured the shadows of the rods in both locations. The one in Syene produced no shadow at high noon on the first day of summer, whereas the one in Alexandria, whose shadow was measured at exactly the same time, made a short shadow (7°12' off the vertical, or one-fiftieth of a great circle). Eratosthenes measured this shadow and compared it with the distance between the two cities, which was estimated as 5,000 stadia. In those days distances were measured by a trained surveyor who walked between two cities counting equal-sized paces. By using these figures—one-fiftieth and 5,000—Eratosthenes was able to determine the circumference of the Earth. His calculations used what was known as a *stadium* or single stadia, which varied according to who did the calculations. It was later determined that an Eratosthenian stadium was 157.5 meters, an Olympic stadium equaled 185 meters, and a Ptolemaic stadium equaled 210 meters. Eratosthenes's calculation for the circumstance of the Earth was later determined to be 25,000 miles; his calculation for the diameter was 8,000 miles. Although his figures were exceptionally accurate, his results were not very well accepted. Most people thought his figures were much too large. (The correct figures for the Earth are 24,902 miles in circumference and 7,926 miles in diameter when measured at the equator, and 24,860 miles in circumference and 7,900 miles in diameter when measured from pole to pole.) Even the fifteenth-century navigators, including Christopher Columbus, calculated the circumference and diameter of the Earth as being significantly smaller than Eratosthenes did. This may be one reason why Columbus's trip to the Americas took much longer than originally calculated.

Centuries later Jean Picard (1620–1682), a French astronomer, used a telescope to measure the distance of a star from its zenith at different places on the Earth; he applied similar mathematical calculations to determine the size of the Earth. His values were very close to our current figures. Sir Isaac Newton (1642–1727) used Picard's figures to confirm his law of gravitation in 1687. The law of gravity states that the attractive force between two bodies is directly proportional to the product of the masses of the two bodies and inversely proportional to the square of the distance between their centers (Gravity: $F = Gm_1m_2/d^2$).

The Mass and Density of the Earth

For many years no one could determine the weight of the Earth because it was so massive there seemed to be no way to weigh it. The law of gravity was proposed to solve the problem, but gravity is a very weak force. You may not think it is weak if you fall out of a tree and hit the ground, but your gravity attracting the Earth is very minor compared with the gravitational pull (mass) of the Earth on you. Henry Cavendish (1731–1810), an English scientist, designed a unique experiment to use the force of gravity to estimate the mass of the Earth. He suspended two small lead balls on each end of a metal rod that was suspended from a thin thread. The rod was free to turn when the slightest force was applied to the balls. He then brought two very large iron balls close to the small balls and the force of gravity between the large and small balls made the rod twist the thread enough for him to measure the amount of twist. Because he knew the distance between the centers of the balls on the rod, and their mass, he was able to calculate the gravitational pull and thus calculate the mass of the Earth. His figure was amazingly accurate. He estimated the mass of the Earth as 3.7 million billion billion pounds, which is very accurate. Once the mass was known, it was easy to determine the density based on the volume of the Earth, which is calculated from the circumference and diameter. The average density (mass/volume) of the Earth is 5.518 grams per cubic centimeter, which is about 5.5 times that of water.

From the beginning of time people speculated about what is inside the Earth. Is it hollow? Are there demons living there? No record exists, but it can be speculated that ancient people thought that the Earth was hollow and that volcanoes and earthquakes were caused by evil spirits. Greek myths say Zeus placed giants, who displeased him, in chains in the center of the Earth. In trying to escape they caused earthquakes. In addition to the Greeks, several religions believed that Hades existed at the center of the Earth because volcanoes produced the fire and brimstone of hell.

A geologist named Athanasius Kircher (1601–1680) published a popular book that stated that dragons lived in many caves inside the Earth. John Cleve Symmes (1742–1814), an American, held the theory that the Earth was hollow and there were openings at the poles where one could enter. The misconception of life in a hollow Earth remains a favorite theme of fiction writers of books and movies. The crust of the Earth (the outer 20 miles) has a density of 2.8 grams per cubic centimeter, which is much less than the Earth's average density of 5.5 g/cc. This tells us that the center of the Earth must be composed of material that is far more dense and heavier than the rocks and dirt of the crust. By measuring the shock waves of earthquakes with a seismograph, we know that the density of the Earth increases with depth. In addition, the internal temperature of the Earth increases with depth. By extrapolating this data it is possible to estimate the temperature at the core of the Earth, which turns out to be somewhere between 9,000°F–10,000°F. Thus, the core is believed to consist of a liquid metal or combination of metals. No one knows for sure because we have not been able to drill more than 6 or 7 miles into the Earth.

The above attributes of the Earth are all requirements for the existence of a carbon-based form of life as we know it. Another attribute needed for life to form was time. We will now consider what we mean by "time" and how our concepts are used to measure the age of the universe, the Earth, and the development of life.

TIME

There are two concepts or meanings of time. One is *temporal time,* which is related to personal, psychological, unmeasurable time. The other is *physical time,* and this is what we are concerned about here. Again, there are two forms of physical time. One is what we think of as *regular time,* which is measurable and always proceeds forward. In other words, it proceeds from the past to the present and into the future. The other is *relative time,* or time related to space as in Einstein's theory of relativity. Relative time can lengthen, shorten, and go forward or backwards. Einstein's fourth dimension of relative time does not really affect our everyday activities on Earth. Here we are dealing with time as related to life; therefore we will concentrate on regular or measurable time, which is linear, cumulative, and can be broken down into units so that we can measure its progress.

Ancient Concepts

Ancient people had only the changing of their environment to consider when they were concerned with time. They could see the regularity of the rhythms of the solar system, the celestial spheres, and the rotation of the Earth—visible as day and night—and thereby conceptualized yesterday, today, and tomorrow. They recognized the macro changes in the seasons, the phases of the moon, and the changes in the positions of the planets. Ancient people also used the stars to help determine longer periods of time, such as months and years. Micro changes, things that occurred in less than a day, were more difficult for the ancients to measure. Their only way to measure smaller units of time was by fractions of a day; for example, dawn, noon, dusk, and night. They had no way to measure time at night. During the day the best way was to break down the movement of the sun's shadow into equal units by placing a stick in the ground and marking off some fractions of time for the day, as indicated by the sun's changing shadow. Originally marks were used for 6 hours of daylight, depending on the season. Thus the sundial was born. In the northern hemisphere the sundial works best if the stick is slanted toward the north pole at an angle equal to the latitude of the placement of the instrument (see Figure 3.4).

The sundial was invented by the ancient Egyptians about 5,000 years ago. The Sumerian's base-12 numbering system was used to divide the sundial's shadows into 12 units to form hours (Greek for "time of day"). It is possible to divide the day (one rotation of the Earth) into units other than 24 hours. The decimal system's base of 10 might have been used; for example, zero could be "high noon" when the sun is at its zenith and "high noon" the next day would equal 100 hours (1 day), which could be divided into 10 or 100 hours, and each hour into 10 or 100 minutes, etc. There is nothing sacred

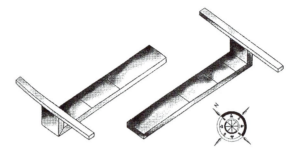

Figure 3.4 Examples of ancient Egyptian "clock" sticks that were used to measure the passage of time during daylight hours. This type of shadow stick was the precursor to the sundial.

about a 24-hour day. Such a metric system for measuring time might be more rational and compatible with other metric units. A variety of measuring systems have been proposed for measuring days, weeks, months, and years. They must all take into account the natural periods of the Earth's rotation on its axis, the Earth revolving around the sun, and the moon revolving around the Earth. Unfortunately, these periods do not equal whole numbers, and they all have minor fluctuations. This complicates time and calendar systems.

The Development of Clocks

There are several ways to measure the passage of time; for example, burning a candle that has marks representing hours, or sand passing through an hour glass. Neither are accurate, because the burn rate of candles varies, as does the rate that sand proceeds from the top to the bottom of hour glasses. The water clock was invented about 1400 B.C. by the Egyptians and Chinese using dripping water. The clock works as follows: water is stored in an upper container, which drips through a small hole at a reasonably steady, controllable rate, into a container below. This clock was refined and improved by the Greeks in about 270 B.C. so that it maintained the same level of water in the upper tank, and thus the same pressure, to ensure an even flow. It was further refined by attaching a floating device that marked off the rise of water in the lower tank and thus the passage of time. The Greeks called their water clocks *clepsydra,* which means "water stealer." Later, weights were substituted for the water, and in 1470 the "spring" was used to drive the clock. This provided the portable timepiece we know now as a watch.

In 1656 Christian Huygens, the Dutch astronomer (see above), applied Galileo's concept of the pendulum to a clock with weights, which provided a regular motion. The regular motion was made more accurate and continuous by the use of an "escapement" movement, which had been invented by the Chinese in 724. The concept of an escapement was to provide mechanical stepwise increments to a toothed wheel, which assured accuracy. An escapement movement is still used today in pendulum/weight and spring-driven clocks. This improved pendulum clock was the first to be accurate enough for scientific use. But it did not work well aboard ships because of the ship's motion. Therefore, a more accurate timepiece was necessary for navigation. In 1728 John Harrison (1693–1776) designed an accurate clock using a balance wheel that would keep time at sea. It was called a *chronometer*. Since that time extremely accurate clocks using radioactive decay rates have been developed. The second is now defined at 9,192,631,770 cycles of the radiation from the radioactive isotope of the element cesium-133. Today many common clocks and watches use small synchronous electric motors timed by the oscillation of quartz crystals.

The Earth's Age

Until about the 1800s, everyone, including scientists, held the belief expressed in the Bible that creation began about 6,000 years ago. This figure was arrived at by several people who worked backward from the time of King Saul, going through each generation (begats) to arrive at this figure. The emperor Constantine counted 3,184 years from Adam to Abraham. St. Augustine used the "begat" system to arrive at the date of creation at 5,500 B.C. Johannes Kepler, the astronomer, arrived at the date 3,993 B.C. Isaac Newton arrived at a date of 3,988 B.C. for the beginning of the universe. And finally, James Ussher (1581–1656), using this technique, calculated the date of creation as nightfall on October 23rd in the year 4,004 B.C. A few years later John Lightfoot (1602–1675) gave his "more exact" date for creation: He said it occurred at exactly 9:00 A.M. on October 26th in the year 4,004 B.C.

By this time scientists recognized that the Earth had gone through changes, but figured 6,000 years was adequate time enough for all the changes such as erosion, the Flood, and so on to take place. Several who questioned this date for creation, the concepts of Adam and Eve, the Flood, and so forth either were burned at the stake or lost their reputations. For example, Bernard Palissy (1510–1589) did not believe in the Flood, but he did believe that the Earth changes over long periods of time. He was burned at the stake for his beliefs. Thomas Burnet (1635–1715) wrote a book that questioned the authenticity of the existence of Adam and Eve. He lost his reputation. Even today, there are many people and religions that hold the belief that the Earth and the beginning of time was about 10,000 or fewer years ago, and that changes were not gradual, but catastrophic. As we shall see, the "catastrophic" theory is again being proposed as an explanation for evolution.

In 1749 Georges-Louis Leclerc de Buffon (1707–1788) used naturalistic terms to write in his encyclopedia that the Earth had to be at least 75,000 years old, and that it became cool enough for life to exist when it was about 40,000 years old. He was forced to recant his statements. Later, James Hutton (1726–1797), an early geologist, in his book, *The Theory of the Earth* (1785), developed a theory called "uniformitarianism." He believed that the Earth had undergone various slow changes that were not necessarily uniform changes. He did not give an estimate of the age of the Earth, but he did state that it was much older than most people believed. This belief was based on his study of rocks and layers of sediment and how long these changes would take to occur. Edmund Halley, mentioned above, calculated the Earth to be at least 1 billion years old based on his study of ocean sedimentation and the salt content of sea water. Thus began the process of determining the age of the Earth more objectively based on physical attributes and less on myths and biblical stories.

Just how do we go about determining the age of the Earth? In the nineteenth century, scientists such as Antoine Henri Becquerel, Marie and Pierre Curie, Frederick Soddy, Ernest Rutherford, Borden Boltwood, and others were discovering and studying new types of elements that gave off radiation and changed from one type of element to another. They were called *radioactive elements* or *radioisotopes of atoms. Radioactivity* was a term coined by Madam Curie. These scientists understood the process of radioactive decay in which radioactive atoms would emit radiation and "decay" into other, simpler types of atoms in sequence (nuclear fission). The final element in the decay process of radioactive atoms was a form of lead that was not radioactive. Once it was understood that this decay is very regular, and has been going on since the beginning of time, scientists recognized the process as a possible time-dating device. Their work led to the concept of *half-life,* a term coined by Rutherford, which indicates the exact time required for half of a radioactive element to decay, leaving the remainder to decay in the next half-life, and so on until the radioactive element was transmuted into a stable element. They could not determine which particular atom or when a particular atom would decay, but they could use averages of the decay of all the atoms, which proceeded at a fixed rate. This process provided an excellent "clock" to use to look back in time by comparing the proportion of radioactive isotopes of uranium and thorium remaining in rocks and sediments with the proportion of these elements that had already decayed into stable lead.

The radioactive isotope of uranium has a half-life of 4.5 billion years, and the isotope of thorium's half-life is 14 billion years. Because they are still found in the Earth and not all of these elements have decayed to lead, it must mean that the Earth is less than 1 trillion years old, but more than 1 billion years old. By comparing the one isotope of lead that is the product of radioactive decay with the decay rate of isotopes of uranium and thorium, it is possible to give an accurate estimate of the age of the Earth. The accepted age of the Earth is 4.6 to 5 billion years, which means it is about 10 billion years younger than the universe.

This brings us to the geologists', rather than the physicists', view of life and the length of its existence. Fossils have always intrigued people. For thousands of years people pondered how these imprints of what seemed to be living forms were found on rocks high on mountain tops. Some of the fossils were obviously similar to clams, sea shells, and other forms of sea life, yet they were found far higher than any sea or body of water. How could this be?

To answer this question we begin with a modern definition of what a fossil is. First, it is evidence of a former living organism or its imprint that is over 10,000 years old. It must be either the original structure or a structure that has been filled with minerals (calcium carbonate or silica from ground water), or imprints or tracks of organisms cast and preserved with minerals. Fossils provide a record of living things going back as far as 3 billion years ago. The study of fossils of prehistoric organisms is called *paleontology.*

Following are some examples of prescientific misconceptions and beliefs about fossils: (a) fossils were left over from the Flood; (b) they existed in the Earth as a result of a natural occult process; (c) Satan placed them there to mislead the faithful and imitate God; (d) God created them for his own purposes; (e) fossils are the results of God "practicing" the creation of living things; and (f) the Bible did not mention them so they do not exist.

Nicolaus Steno (1638–1686), a Danish geologist, was one of the first scientists to state that fossils were the remains of very old living things whose remains became petrified.

We now know that buried in layers of sedimentary material there is a record of living organisms that are approximately 3.5 billion years old. We also know that the deeper layers contain the oldest fossils. This helps scientists arrive at the various ages of the Earth that are called periods, eras, and epochs and the progression of different species of life forms that accompanied these periods.

There are two basic methods of dating that scientists use to determine the age of rocks and fossils. One is the *relative* method, the other is a series of *absolute* dating methods.

The *relative* method of dating the age of the Earth and fossils is based on geology. Relative dating is based on the principle of stratigraphy. It is assumed that in an undisturbed series of strata of rocks, the oldest layers are on the bottom and the youngest are on top. This system was used to develop the scale of geological history, which is divided into four major eras— Precambrian, Paleozoic, Mesozoic, and Cenozoic. These will be discussed in the section on evolution.

The development of absolute dating methods provides more accurate and definitive scales. Radiometric dating is based on the regular decay rates of radioactive elements. These steady half-life decay rates provide an accurate "clock" of the age of the rocks in which radioactive elements and their stable end isotopes are found. Because uranium-238 has a half-life of 4.5 billion years it can be used to accurately determine that Earth is about 4.5 or 5 billion years old.

The carbon-14 dating method is based on the half-life of carbon-14 being 5,730 years. Carbon-14 dating is ideal for dating fossils of living things, but it only goes back as far as 50,000 to 70,000 years. It is useful for dating organic materials because all living things contain some carbon, both carbon-12 and radioactive carbon-14.

The rubidium–strontium method is used to date old Earth and moon rocks. It is based on the beta decay of rubidium-87 into the element strontium-87. A similar method involves thorium-230, which is used to date old ocean-floor sediments. Another method uses lead-alpha radiation to help establish the geologic age of different strata in the Earth.

There are a few absolute dating methods that do not make use of the decay of rates of radioisotopes. One is called *dendrachronology*, which involves examining the rings in old trees that are cut cross-wise. The shapes and sizes of the rings give clues to recent past climate and environmental conditions up to about 3,000–4,000 years ago. The *obsidian hydration dating method* uses the water "arks" or artifacts made of obsidian (volcanic glass). It is useful for dating up to 200,000 years ago. Another method, *thermoluminescence,* is used by heating ancient pottery until it glows and then measuring the electrons freed in the minerals of the clay. It can be used to date pottery back several hundred thousand years. Still another method measures the sedimentary beds left over from glaciers during the Ice Age (Pleistocene epoch). It compares the rate of sediments deposited over the years as compared to geologic events.

EVOLUTION

For hundreds of years there has been speculation, confusion, and misconceptions regarding evolution. In fact, many people did not believe that evolution was and is a probable phenomenon, or even that there was much evidence to support the concepts of organic evolution, particularly when these concepts are applied to humans. That disbelief is still prevalent today among some groups of people, but things are changing. In October of 1996, Francis D'Emilio wrote an article entitled "Pope Claims Science, Faith Can Coexist." D'Emilio continues:

In his most comprehensive statement yet on evolution, Pope John Paul II insisted that faith and science can coexist, telling scientists that Darwin's theories are sound as long as they take into account that creation was the work of God. . . . In a statement released Wednesday, the Pope said new knowledge has confirmed that Charles Darwin's theory of evolution is "more than a hypothesis." . . . his latest comments were the clearest and most comprehensive in support of Darwin's conclusions. . . . the Vatican places the Roman Catholic viewpoint in stark contrast to that of some fundamentalist Christians, who believe the biblical account of Creation is literal.

The modern concept of organic evolution is referred to as *neo-Darwinism,* which combines Mendelian genetics with Darwin's concepts of evolution. Essentially, evolution is the adaptation of organisms to changing environments. The organisms best adapted to these changes have the greatest number of offspring and thus out-populate those who do not survive. This is referred to as *natural selection,* which ensures the passage of successful genes and characteristics required for continuation of the species. The arguments concerning scientific evolution and religious creationism continue today among scientists, theologians, laypersons, local governments, and school boards.

Many of the misconceptions about evolution and religion are based on a misunderstanding and lack of knowledge of just what evolution is all about. Some people "fear" learning something that they mistakenly believe conflicts with their religious or political ideologies.

The most direct proof of evolution, that is, the progression of primitive life forms to more "advanced" plants and animals exists in the study of the past through fossil remains. *Fossil* is the Latin word for "something dug up." Paleontology is the study of prehistoric organisms through the analysis of fossil remains.

As previously discussed, paleontologists and geologists can determine both the age of the rocks and the age of fossils contained in them. Using a technique called *trace fossils,* the development and changes in plants and animals, as well as their age, and the period during which they existed can be traced. Most of the billions of simple plants and animals that lived billions of years ago were soft bodied and decayed on death. Therefore they left no trace. But billions more were trapped in mud, which was eventually overlaid with tons of sediment, until their imprints were molded in the mud/rock. Finally, a "cast" was formed with silica and calcium carbonate deposits. In some situations we are left with imprint molds of footprints or soft parts of plants and animals, such as wings, feathers, leaves, and leaf veins. Fossils are also composed of minerals that replaced the shells, bones, and teeth of animals, and the wood and bark of plants buried in sediment layers.

Catastrophism

Several theories of evolution are still debated today. The first, *catastrophism,* was proposed in 1812 by the anatomist, Georges Cuvier (1769–1832). His theory was that there were many "creation periods" and each one ended in a catastrophe, which was followed by new creation of life. Each time, the new life was more and more like present-day life. Cuvier is considered the father of paleontology. His main misconception related to his concept of gradual evolution.

Catastrophism led to a more modern concept called *punctuated equilibria* or *punctuated evolution,* which was proposed in 1972 by the paleontologists Stephen Jay Gould and Niles Eldredge. Their theory states that most species did not change and were stable for long periods of time until a drastic environmental change, such as the impact of a comet or meteor, or some change in food supplies, or change in the Earth's temperature occurred. Following these periods of drastic change, many species of plants and animals died, but a small group of some species made drastic changes and developed into new species more suited for the new environment. This was explained as a stable period, punctuated by occasional periods of rapid evolutionary change. Punctuated evolution is not yet accepted by most scientists. Most biologists have no doubt that biological evolution has and is taking place.

Gradualism

The second theory of evolution involves the concept of *gradualism.* One of the first scientists to write about evolution was Georges-Louis Leclerc de Buffon (1707–1788). His misconception was called "degeneration," which meant that lower life forms degenerated from higher forms; for example, apes degenerated from humans, donkeys degenerated from horses, small insects from larger insects, and so forth. His belief was somewhat the reverse of accepted gradualism. But at least he was considering changes over long periods of time.

Creationism

The Canadian geologist Sir John William Dawson (1820–1899) did not accept or believe in Darwin's theories of evolution. Some people in both Canada and the United States used Dawson's position as an authority to promote a more literal interpretation of the Bible. This relates to a third theory of evolution, called *creationism,* which is a religious concept of evolution. Creationism became a political issue in the twentieth century. It has been substituted as a "science" to be taught in some schools by political bodies, who were influenced by religious groups. Creationism has been such a contentious misconception that, even though it is not based on scientific evidence, some school districts still insist that it be taught along with the science of evolution. Originally, creationism meant that God created the soul of each human at the time of his or her birth. Now it means a literal translation of Genesis as to the origins of the universe, Earth, and life, including the creation of humans about 10,000 years ago. Although creationism is not accepted by mainstream science, there is a point to be made about the concept. If you go back far enough into the development of life from the organization of simple atoms and molecules into "networks" of autocatalytic molecules that can reproduce, you can conceive of a metaphysical "creation" at the point that this prebiotic "life" formed and evolved over eons of time.

Other Concepts of Evolution

Other concepts of evolution and the origins of life are being explored by several scientists at the Santa Fe Institute in New Mexico, as well as at other organizations and universities. This new approach is concerned with complex dynamic systems of all types, not just life and evolution. They are developing algorithmic computer programs and models based on mathematics and the chemical, physical, biological, and social sciences. They are exploring exciting new insights into areas such as chaos theory, complexity, spontaneous self-organization, order from disorder and vice versa, autocatalytic networks of molecules, coevolution, the roles of cooperation and competition in the evolution of complex social and biological systems (including economics), and especially, models of artificial life. They approach the origin of life and evolution from the development of the most simple chemical and physical organizations to the point of "life," and then move to the more complex. This is just the opposite of some older approaches to evolution that started with humans and proceeded to the more simple organisms.

Figure 3.5 shows some of the thinking of the Santa Fe Institute scientists' concerns about complexity, life, and evolution.

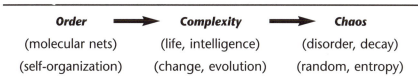

DYNAMIC SYSTEM

Order	→	*Complexity*	→	*Chaos*
(molecular nets)		(life, intelligence)		(disorder, decay)
(self-organization)		(change, evolution)		(random, entropy)

Figure 3.5 A dynamic system for evolution.

The history of evolution has passed through several phases or developments. Today it seems that some of these theories are being recycled.

Even before Buffon came Pierre Louis Moreau de Maupertuis (1698–1759), the first individual to propose a general theory of evolution similar to what is accepted today. Although he did not know about genes, he suggested that "particles" of hereditary matter were passed from one generation to the next.

Carolus Linnaeus (1707–1778) believed in Genesis, not evolution, but by developing a system of *taxonomy* (from the Greek word for "naming in order") he made a great contribution to evolutionary theory nonetheless. His classification system was originally devised for plants but was later applied to the animal kingdom. Linnaeus used the differences and similarities in organisms to classify them into *species,* which as similar groups formed *genera;* similar groups of genera were named *classes,* and these were placed in large groupings called *orders.* In other words, the group of plants and animals most like each other were the "species," which were then combined into larger groupings, and then larger groups, and so on. He used a tree with its trunk and major branches as an analogy. Even though he did not accept evolution, his taxonomy made the concepts of evolution more reasonable and acceptable to other scientists.

Jean-Baptiste de Lamarck (1744–1829) continued with Linnaeus' classification and invented the terms *invertebrate* (animals without backbones) and *vertebrate* (animals with backbones). His big misconception about evolution was based on the theory of "use and disuse" of organs and how this would affect the inherited characteristics of offspring. He felt that one could acquire specific traits and these could be incorporated into the hereditary material that was passed to the next generation. Much later the science of genetics disproved his theory of the inheritability of acquired characteristics.

Charles Darwin

There were others who made contributions to evolution, but none more important than Charles Robert Darwin (1809–1882) and Alfred Russel Wallace (1823–1913), who independently theorized natural selection at about the same time. Even though Darwin did much of the work formulating his theory of evolution early in life, he delayed publication.

At the age of 22, Darwin sailed on the HMS *Beagle,* where at various island ports he observed and collected plants and animals, which he studied to lay the foundation for his work. He had already read about Thomas Malthus, who proposed that environmental factors (famine, diseases, and climatic changes) can limit populations. Malthus also had some vague ideas about natural selection based on his population theories. It was not until

1859 that Darwin published his famous book, *On the Origin of Species by Means of Natural Selection or the Preservation of Favored Races in the Struggle for Life,* which is often spoken of as the "book that shook the world." His theory, in essence, is that because of the food supply and other changes in the environment, plants and animals compete for survival. He also believed that natural forces continually changed the environment in which plants and animals lived. These organisms that survive produce the next generation, which may thrive because of inherited favorable traits, which again may be passed to the next generation. Darwin believed in a form of Lamarckism, because he had no knowledge of genetic's role (DNA) in inheritance.

Darwin's great insight was that an "agent" can improve the model of a living organism without the intervention of any paranormal (religious) interference. All the system (organism) had to do was try out models to see how well they worked in the environment. If the new model survives, then it will adjust the existing model to be even more successful in future generations. Darwin also stated that all plants or animals that are related to each other were derived from common ancestors. His concept of variations in individuals being passed to future generations was not proven until Mendel's Law was accepted, which led to the birth of modern genetics. Darwin was the first to consider evolution as progressing from simpler organisms into more complex species, even though he did not know about the mutation of genes. Contrary to popular belief, Darwin did not use the phrase "survival of the fittest." This concept states that organisms that are less well adapted to their environment tend to die, whereas those that are better adapted survive. The phrase was actually coined by Herbert Spencer (1820–1903). Darwin's concept was related to natural selection as the process for evolutionary change.

More recently, some scientists have combined early hereditary (Mendelian genetics) with Darwin's concepts of evolution. This new, mechanistic branch of evolution is referred as *neo-Darwinism,* which results from the combination of random inherited changes or mutations through the genes of individuals.

Autopoiesis

Some scientists consider neo-Darwinism "mechanistic" and have proposed a new physiological concept for evolution called *autopoiesis.* Autopoiesis is a term proposed by Humberto Maturana to define an interrelated, self-maintained, carbon-based chemical system of life, rather than a mechanistic system. It is derived from the Greek words *auto* (for "self") and *poiesis* (for "to make"). Autopoiesis is composed of six physiological principles that are functions of living organisms, including metabolism and chemical/physical functions. According to Margulis and Sagan (1992), the six evolution-ary functions of autopoietic systems are:

1. *Identity:* There is organization of the internal contents of cells—thus internal components are identifiable and bounded by a containing (cell) membrane. Inside these boundaries living systems contain nucleic acids, proteins, fatty acids, and so forth.

2. *Integrity:* Although a single, unitary system, an autopoietic system is a dynamic functioning, multienzymed network of acids and proteins.

3. *Self-boundedness:* A membrane or cellulose-type cell wall is produced as a bound-ary structure by the system.

4. *Self-maintenance:* Within the system's boundary, carbohydrates and other types of synthesis are produced by the functioning system.

5. *Raw material:* An external supply of elements such as H, C, N, O, S, and P as well as enzymes are incorporated into the system. They are required for metab-olism.

6. *External energy:* Light is the main source of energy required to convert chemi-cals into energy, through bonding. Chlorophyll and bacteria combine sugars and other organic compounds in the system.

Autopoietic systems rely on metabolism (resources and energy) to main-tain the systems. In other words, its views of evolution are more "Gaia-like," and are based more on the physiology of living organisms than on the anatomic structure of plants and animals, which is often expressed as "func-tional determinism" in classical Darwinism.

Lysenko

Lamarck's views on the inheritance of acquired characteristics were mild and understandable within his time (see "Other Concepts of Evolution" in this chapter). But later, in the early twentieth century, Trofim Donisovich Lysenko (1898–1976), who expanded Lamarck's incorrect theory of the inheritability of acquired characteristics, destroyed a whole generation of biology and genetics in his home country, the USSR. With Stalin's approval, all scientists who opposed Lysenko were either killed, sent to Siberia, or escaped to the West. By the middle of this century all scientists in other countries rejected both Lamarckism and Lysenko's views. But the Soviet Union thought that the concept of passing on inherited characteristics from generation to generation fit neatly with Marxist theory. Therefore Lysenko was promoted by Stalin, as were his theories.

Lysenko rejected the concept of genetic mutation. He postulated that by stretching their necks to eat leaves higher up on trees, giraffes would survive to pass on the acquired long necks because they could get the most food, and they would then pass on this acquired characteristic to offspring. As head of both the Institute of Genetics and The Soviet Academy of Science, he conducted many plant-breeding experiments which he said proved his theory and would improve agriculture worldwide. Unfortunately, Soviet agriculture did not progress, so in the middle of the twentieth century, after Stalin was gone, Lysenkoism followed.

Many people (and some religions) believe in the misconception that it is possible to inherit acquired characteristics. Many parents still hold the misconception that if they have developed positive habits and traits in their lives that these acquired traits will be passed on to their offspring genetically. This hasn't been proven through science and genetics. It would make just as much sense to believe that if a father cuts his little finger off of both hands, all his children and grandchildren will be born without pinkies.

Lysenkoism is an excellent example of a situation in which accepted scientific truth is subordinate to political control. A similar twisting of science for political purposes occurred in Nazi Germany in the 1930s and 1940s.

There are still some arguments as to just what drives the processes responsible for evolution. What biological, chemical, physical, and environmental mechanisms are required for organisms to evolve over time?

Genetics

Many scientists from the late Middle Ages until the late nineteenth century speculated about some type of "unit," or as Darwin called it, an "agent," that could carry inherited characteristics from parents to offspring. This was before people knew about chromosomes, genes, DNA, or even before the term *genetics,* from the Greek, meaning "give birth to," was used. Genetics may be defined as the study of heredity and the changes that affect inherited characteristics. We have already mentioned probably the oldest and most persistent misconception about genetics. Namely, that acquired characteristics are inheritable. But how did genetics change this misconception?

Gregor Mendel (1822–1884) is considered the founder of genetics. He used statistical methods to analyze the results of cross-breeding pea plants to determine the ratios for the appearance of different characteristics in future generations. He did not actually see the genes, chromosomes, RNA, and DNA as we know them today. He was able to deduce what a gene must be like in order to determine the 3:1 ratio of dominant to recessive hereditary factors his work demonstrated. Mendel's plant-breeding experiments did not result in an exact 3:1 ratio of selected characteristics, but his insight led to the use of statistics to arrive at this deduction. And, from this, he worked out the Mendelian Laws of Inheritance.

It was not until 1882 that Eduard Adolf Strasburger (1844–1912) observed the division of plant cells through a microscope. He is credited with selecting the names *nucleoplasm* for the protoplasm inside the nucleus, and *cytoplasm* ("cyto" means cell) for the protoplasm outside the nucleus. Several years later, Walter Flemming (1843–1905) used dyes to color parts of the nucleus. He called the material inside the nucleus *chromatin,* from the Greek word for color. He soon realized that this chromatin was a short threadlike material that would divide. These structures were called *chromosomes,* meaning "colored bodies" in Greek. Unfortunately, Flemming did not draw any conclusions about the role of chromosomes in the inheritance of biological traits.

Thomas Hunt Morgan (1866–1945) was familiar with Mendels's work with plants. He wanted to verify Mendels's Law of Heredity with animals. Morgan selected the tiny fruit fly for study because it has only four chromosome pairs in each cell, and reproduced many generations in a short period of time. Humans have 22 pairs of autosomal chromosomes (nonsex), plus 1 pair of chromosomes that determine the gender of the offspring. Morgan discovered that specific characteristics were inherited from one generation of fruit flies to the next. He also noted that parts of the chromosomes interchanged with each other, but he could not identify the individual units.

In 1909 Wilhelm Ludwig Johannsen (1857–1927) suggested that the parts of the chromosome that carried the inherited characteristics be named *genes.* Later, two forms of genetic material were identified as being responsible for heredity: RNA (ribonucleic acid) consists of a single chain of genetic material that carries information, whereas DNA (deoxyribonucleic acid) is a double chain that forms a double helix of repetitive building blocks of inheritable genetic material.

The history of RNA and DNA goes back some years and is related to the work of several scientists. In 1869 Johann Friedrich Miescher (1844–1895) discovered a new substance in pus that contained both nitrogen and phosphorus. His friend, Felix Hoppe-Seyler (1825–1895), identified the substance as similar to yeast and determined that it originated in the nuclei of cells, so he named it *nucleic acid.* Albrecht Kossel (1853–1927) was able to isolate the nitrogen base of nucleic acid, and later, in 1909, Phoebus Aaron Theodore Levene (1869–1940) identified sugar in nucleic acid and identified it as "ribose," which came to be known as ribose nucleic acid, or RNA. Later, Levene identified another sugar in nucleic acid, which was called deoxyribose because it was missing an oxygen atom in the sugar molecule. Thus, the second type of nucleic acid was named deoxyribonucleic acid or DNA.

DNA is the basis of inheritable characteristics found in the nuclei of the sex cells of both parents. If the DNA contained in a single human cell were laid out end to end, it would be about 6.5 feet long. If all the DNA in the trillions of cells in a single adult human were laid out end to end, it would extend almost 20 miles. DNA molecules are chains of protein nucleotides of only four different kinds: adenine (A), thymine (T), guanine (G), and cytosine (C). All living things contain these four hereditary building-blocks in their cell nuclei. An adult human's body contains over 1000 million million cells. All the nuclei of these cells for a single individual contain the same specific DNA code. While over 95% of all DNA is the same in all species, the sequence of the building blocks in humans is different from other species. Additionally, one human's DNA sequence is not the same as another person's sequence (with the exception of identical twins). Based on the work and background research of many scientists, Francis Harry Compton Crick (1916–) and James Dewey Watson (1928–) were able to unravel the complex structure of DNA. They used X-ray diffraction photographs taken by Rosalind Elsie Franklin (1920–1958) as the basis of their study. There is some controversy over whether Crick and Watson had permission to use Franklin's DNA photographs. Franklin, from her research, suggested that the DNA structure was similar to a spiral staircase. For their work in determining the complicated double-helix structure of DNA, Crick, Watson, and Wilkins (Franklin's boss) shared a Nobel Price in 1962, but as Franklin was no longer living, she received no credit.

Even after the information about the roles of RNA, DNA, and genes in genetics became available, many people still had (and have) misconceptions about what can and cannot be inherited. The role of genes, usually in some combination, determines almost all of our physical and psychological characteristics. Many people believe that our experiences and environment are more important than what we inherit from ancestors. For instance, a major misconception is that intelligence is based on *nurture* or what we are taught, and that *nature* is of little importance. Scientists have tried to determine to what degree genetics and environmental factors each contribute to our intelligence. No answers are firm, but we do know that a person's genetic makeup does determine his or her capacity for learning and intelligence. Presumably, a person's ultimate intellectual capacity or potential has never been reached or measured.

The role of genetics and DNA in life and evolution is significant. At times, things just do not proceed according to plan, and parts of the genes and DNA do not match up as they should. These "mistakes" are referred to as *mutations,* which are what drives biological evolution.

Hugo Marie De Vries (1848–1935), who was familiar with Mendel's work with pea plants and the laws of heredity, noticed that his primrose plants did not always breed true to type. At times new characteristics appeared in the plants that De Vries had never seen before. This led him to surmise that evolution was not always a slow process, that at times evolutionary changes could be rapid. He called these visible changes *mutations,* from the Latin word for change. In 1937 Theodosius Dobzhansky (1900–1975) combined the field of genetics with the concepts of naturalists. Through the use of mathematics and experimental techniques he combined mutation with evolution. Ever since, evolution has been understood to take place at the molecular level within the cell's nuclei, genes, and DNA.

THE EVOLUTION OF HUMANS

Paleontologists divide the past 570 million years of the Earth's history into *eras, periods,* and *epochs.* The *era* is the longest major geological division of time. Each era is subdivided into periods and epochs. There is evidence of microscopic life forms existing as long as 3.5 billion years ago. The Precambrian fossils were very small and had soft bodies so there is not much fossil evidence of them. The first fossils of larger animals with shells and bones formed about 600 million years ago.

These periods, their approximate age, and some typical examples of the types of fossils found in each geological time period follow:

Precambrian Era: (Age of Beginning)

Period

Azoic	4600+ mya*	Origin of the Earth
Proterozoic	3800+ mya	Bacteria, primitive algae, fungi, protozoans

The Paleozoic Era: (Age of Ancient Life)

Period

Cambrian	570 mya	Algae dominant
Ordovician	505 mya	First land plants, first fish, and invertebrates
Silurian	438 mya	Algae dominates, first insects, and arachnids
Devonian	410 mya	Land plants, forest, gymnosperms, fish, amphibians, millipedes appear
Carboniferous	350 mya	Fern forests, swamps, ancient sharks, mollusks, first reptiles
Permian	285 mya	Conifers (pines) evolve, modern insects, mammal-like reptiles appear

The Mesozoic Era: (Age of Reptiles)

Period

Triassic	245 mya	Gymnosperms and ferns, first dinosaurs, egg-laying mammals
Jurassic	208 mya	Ferns and gymnosperms, large, specialized dinosaurs, marsupials
Cretaceous	135 mya	Angiosperms appear, first modern birds, dinosaurs peak and begins extinction

The Cenozoic Era: (Age of Mammals)

Period

Tertiary

Paleocene epoch	65 mya	Dinosaurs extinct, evolution of mammals
Eocene epoch	58 mya	Beginning of age of small mammals, modern birds
Oligocene epoch	37 mya	Flowering plants, forests, modern mammals appear, primates, apes appear, big cats
Miocene epoch	25 mya	Modern nonhuman mammals evolve
Pliocene epoch	5 mya	Grass lands, flowering plants, grazing mammals, first human-like primates

Quaternary

Pleistocene epoch	1.6 mya	Four ice ages, extinction of many plants and mammals, early man appears
Holocene epoch	10,000 yrs ago	Decline of (modern) woody plants, rise of herbaceous plants, age of *sapiens,* modern humans dominate

* Note: "mya" = millions of years ago.

This breakdown of specific evolutionary periods places human development within the natural order. Of course, one major misconception is that humans did not (and do not) evolve—but rather they were created relatively recently, essentially in our current form or state, and have been much the same ever since. This basic concept of the origin of humans, with some minor differences, has existed in almost every culture and persists today. Another major misconception is that humans evolved from apes.

The current theory is that modern man (*Homo sapiens*) evolved from *Homo habilis*. About 5 million years ago modern humans split off from *Australopithecus* (ancient primatelike man) to form the genus *Homo* (abbreviated as *H.*). *Australopithecus* fossils have been found in south and east Africa. The *Australopithecus afranesis* was found in Ethiopia by Dan Johanson in 1974. The skeleton was of a 3.5-foot female, about 25 years of age, who was given the name Lucy (after the Beatles's song, "Lucy in the Sky with Diamonds"). Lucy is estimated to be about 3.3–3.5 million years old. *H. habilis* evolved about 2.5 million years ago. About 1.5 million years ago *H. erectus* (walks upright) evolved, most likely in southern Africa, or possibly Asia. It was not until about 200,000 to 300,000 years ago that *H. erectus* evolved into *H. sapiens,* who was not as advanced as modern man. Modern man is called *H. sapiens sapiens* (wise, wise man) and appeared only about 10,000 to 50,000? years ago based on fossil remains. There is no agreement as to whether the evolution of humans was smooth or if there were some major interruptions in the process. There is disagreement as to the place, time, and roles of Neanderthals and Cro-Magnons (found in Europe 300,000 to 30,000 years ago) with regard to human development.

The DNA evidence and comparisons of blood proteins indicate that modern man did not evolve from apes, but rather split off from a common ancient ancestor, most likely by genetic mutation, about 6–8 million years ago. The oldest humanoid fossils found so far are about 5 million years old, so the best guess of the time our human ancestors developed into their own branch of bipeds must be between 5 and 6 million years ago. It is obvious that it has taken humans a long time to arrive at what we consider the beginning of recorded history.

The misconception that humans evolved from apes was held by many people who most likely believed in Buffon's misconception of degeneration, in which "higher" animals and humans degenerated into lower animals. Once Darwin's theory of organic evolution became known many people just assumed that now scientists were saying that higher animals were generated from lower animals, and humans evolved from apes.

Many people still adhere to the misconception that organic evolution tries to show that humans evolved from apes. Only those who have not read *On the Origin of Species* (Darwin, 1859) or who misunderstand organic evolution, still consider humans *direct* descendants of apes. Darwin did not say humans evolved directly from apes, but even so many people still consider organic evolution antireligious. Darwin did relate human vestigial or rudimentary organs and their functions to subhuman species. For instance, the coccyx or tailbone remnant in humans is assumed to be a nonfunctioning tail left over from animal tails, just as the human ear muscles, nictating membrane (inner eyelid), and appendix no longer function in humans. Early fossil remains that might indicate the branching of humanoids from apes have not yet been found; this is the so-called "missing link." We do not yet have information that gives us exact knowledge as to the earliest evolution of humans. The latest theory is that over many millions of years all life-forms on Earth evolved from very simple celled and multicelled organisms.

Chapter 4

Chemistry and Physics

INTRODUCTION

As biological life has a chemical basis, so does chemistry have a physical basis that drives the chemical reactions of atoms and molecules as they form more complex compounds and substances. When one begins examining matter and energy in the universe, the understanding of the physical nature of everything becomes evident. In this chapter we will trace scientific developments, beliefs, and misconceptions related to chemistry and physics that people have held for thousands of years.

We begin by exploring some of the ideas proposed by ancient people as they tried to comprehend the world around them. We then proceed to the Greek/Roman period, the Dark and Middle Ages, the Renaissance or Enlightenment, and end with some current scientific developments, beliefs, and misconceptions of the twentieth century, leading into the twenty-first century.

ANCIENT DEVELOPMENT OF THE USES OF FIRE AND TOOLS

Since the beginning of time, men and women made use of all types of chemicals, physical reactions, energy, engineering, and even technology without developing a systematic body of knowledge to explain and understand what was involved. Surely, when ancient people wanted to move a heavy log or boulder, someone communicated the idea of placing sticks under it to pry it up and make it move. Thus, the lever was used without understanding the mechanics involved. Many similar examples can be found. But most science must have begun by men and women making observations, communicating with each other, and then making use of all kinds of chemical and physical laws without understanding more than the fact that they could get things done more easily or improve their chances of survival.

Fire is both a chemical and physical process. It involves changes of matter and it involves energy. The earliest use of fire was near the end of the last major Ice Age, about 500,000 years ago. Ancient humans (*Homo erectus*) most likely were the first to use fire, possibly caused by lightning. They learned to contain and control fire, which is one of the characteristics that set prehistoric people apart from animals. Almost all other animals feared fire as they escaped from burning forests and grasslands. At first, ancient people learned how to use fire for warmth as they moved north during the time when glaciers were receding. They also learned how to use fire as protection from animals. They used it to fashion pointed sticks that made better weapons and tools. Fire produced a chemical reaction in the wood that charred and hardened the wood fibers. Ancient people also learned to cook meat, which made it tender and easier to eat. Later, fire was used to bake clay to form pots that could hold water. Because ceramic pots were not good heat conductors, they soon learned to place hot stones in the water inside the pots, which would heat the water and cook their food.

As far back as 2 million years ago, an even older species of humans (*Homo habilis*) used clubs of wood and bones and, most likely, threw stones as weapons for hunting. By the time *Homo habilis* disappeared from the scene, *Homo erectus* was the only species of human left in the world. *H. erectus* soon learned how to design and use more advanced tools and weapons; for example, they fastened a sharpened stone to a piece of wood using strips of wet hide, which would shrink while drying and tightly hold the stone to the wood handle. The mechanics or "physics" of swinging an axe with a heavy stone on one end increased its killing power (mass + circular velocity = angular momentum).

The results of this physics were not lost on the hunters and warriors of early times. The hand axe was a very successful technological instrument by any standard. People of many countries used and improved battle axes over thousands of years. The ax was only supplanted in the logging business a few decades ago with the invention of the two-man saw, followed by the gasoline chain saw.

Similar physical principles were used to increase the deadliness of sling shots. A stone was placed in the center of a long leather strap. By holding the ends together and rotating it rapidly over one's head, the angular momentum and centrifugal force were increased. One end of the strap was released, sending the stone with deadly force, but little accuracy. Later, people learned how to make a hinged device, called a spear-thrower, that could release a spear at an accelerated velocity much greater than one could achieve by just throwing the spear. Someone also figured out how to fashion a string of hide or twisted fibers to a bow of wood that would send a small spear (arrow) at an even faster speed, further, and with more deadly force. Bows and arrows were in use sometime after 25,000 B.C. The increase in the speed (velocity), times the weight (mass), resulted in greater momentum. Thus, the increase in killing power was recognized, but the physics involved were not understood for several thousands of years. These are examples of technological developments, not science as we know it.

In addition to fire, stone tools, and weapons, there are other early uses of chemistry and physics that enabled humans to more successfully deal with nature. A few examples are the use of wood fires to smelt metals from their ores, using the chemical reaction called reduction. This led to the production of copper (about 4,000 B.C.), then brass and bronze alloys (about 3,600 B.C.) and finally iron tools and weapons. Iron was first separated from its ore by Armenians around 2,000 B.C. Iron was more difficult to obtain from ore because it required using charcoal to produce greater heat than that needed to smelt copper, tin, zinc, or lead.

Friction and the Wheel

People struggled for centuries to overcome a problem of physics that still faces us today—friction. At first, heavy objects had to be pushed and pulled over the ground—the smoother the ground the easier it became to move large stones. Early people also discovered that wetting the ground made the job easier—even though they did not understand the physics of friction. (Friction is the resistance to the tangential motion [force] of one body in contact with another.) The greater the imperfections of the sliding surfaces, their masses, and the speed of movement, the greater the friction. The problem was how to reduce this resistance. Later, possibly as early as 5,000 B.C., logs were used as rollers, but friction was still a problem. It was not until about 3,500 B.C. that people came up with a unique way to reduce friction even more—they "invented" the wheel, possibly after inferring a wheel from the turning table used by potters to make round clay pots. Wheels were used on carts to reduce friction even more. They did not understand the physical concepts of friction, energy, force, and momentum, but this did not hinder them in developing technologies to lighten their burdens.

The Chemical Process of Ancient Winemaking

In 1996 researcher Patrick McGovern reported the discovery of winemaking residue in the bottom of a 7,000-year-old clay pot found in the Zagros Mountains of Iran. This extended the history of the chemistry of winemaking (fermentation of sugars by yeasts producing CO_2 and alcohol) by 2,000 or 3,000 years. The ability to make wine existed, even if the exact chemistry of the process was not understood for several thousands of years. In fact, it became a well-known process and was easy for anyone to use. By the year 2000 B.C. there were laws holding people responsible for their drunkenness. Another use of the fermentation process began some years later. It was discovered that if a mixture of flour, fat, honey, and water was left open to the yeast in the air or if yeast was added before the dough was baked, it would rise. After rising, because of the production of CO_2 gas, when the dough was baked a "spongy," more palatable bread was the result (the yeast was eliminated by the heat).

Early Technological Developments

Because ancient people did not understand the chemical and physical basis of life, they developed explanations that must have seemed rational to them. Concepts of the "unknown" emerged as mysticism and religious systems. An important factor in the increase in people's development of the sciences and technology, as well as religions, was the development of wealth, which resulted from more settled communities that brought job specialization, agriculture, animal husbandry, and writing. Sometime after 2,000 B.C. elements of geometry and algebra were developed to assist astronomical calculations and building of temples and pyramids. As travel methods improved, knowledge about medicine, engineering, mathematics, metallurgy, as well as chemistry in the form of alchemy was passed from country to country as merchants visited distant markets and armies invaded different countries.

Trial and error, often based on insight, was how most technological devices were developed. But some of the reasons why these devices "worked" were based on misconceptions that resulted from a lack of knowledge of chemistry and physics. Trial and error played an important role in the early development of agriculture, medicine, construction, and transportation. The development of boats is a good example. No doubt primitive people observed that logs floated, so they used logs as boats. Along the way someone noticed that it made more sense to propel the boat with a branch rather than their hands, so the paddle was invented. Then someone decided to improve the stability of the log by tying branches crossways to the log, and later to tie logs together to make a raft. A next step may have been to hollow out the log with fire and axes to make a boat. These, and later developments were based on everyday experiences, not knowledge of buoyancy and the related physical concept of displacement, which had to wait for Archimedes.

Early Development of the Steam Engine

Another example is the development of the "steam engine" by the ancient Greek, Hero (also spelled Heron). He made a hollow ball with two bent tubes protruding from the sides. When he boiled water inside this device, steam came out the bent tube and turned the ball, which was suspended over fire. Hero, and others, until the time of Isaac Newton, incorrectly believed that the steam was pushing against the air, thus making the ball spin backward from the steam. With his third law of motion, Newton explained that the reaction of the escaping steam caused a reaction in the opposite direction inside the ball, thus causing the ball to revolve away from the escaping steam. Unfortunately, Hero never developed his device for any practical use so he is not usually considered the inventor of the steam engine.

The Smelting of Metals

Over the ages people held many false beliefs about chemistry, particularly the use of chemicals for the smelting of metals from their ores. Early miners held the misconception that difficulties in extracting some metals from their ores were caused by evil spirits. Consider the example of ancient miners who could not get cobalt from some copper ore. They called the ore *kupfernickel,* which means "the devil." Later, a new metal was extracted from this ore and it was named *nickel,* after the name the miners gave to the mysterious "Old Nick's Copper."

FOUNDATIONS FOR THE SCIENCES OF CHEMISTRY AND PHYSICS

We will explore some of the history of the beliefs and misconceptions held by the Greeks and later the Romans, who for many centuries relied on myths and legends for answers to their questions. Sometime after 700 B.C. to about the first century A.D., many Greek myths were reworked into philosophy. During this period the Ionian philosophers were looking at nature more impersonally than did former generations, who considered mythical gods and legends very personal. This was also the period of the decline of the Greek city–states and the rise of the Roman Empire. Since this period of history, classical Greek/Roman philosophy has influenced thought, particularly in the Western world.

Matter

From this period and for the next 2,000 years, philosopher/scientists used the word *material* to mean matter, which was any substance that could be measured with three dimensions (height, width, and depth), and had characteristics such as weight, impenetrability, color, odor, texture, and so forth. Today, we usually think of such matter as *chemical substances.* (Because the early scientists could not see or measure most gases, they more or less ignored gas as a state of matter.) According to Aristotle, until the *material* of something was known there could be no understanding of the *formal* or "form" of the object; for example, the way things behaved in nature. They also used terms like "space," "emptiness," "infinity," "forces," "motion," and so on with little understanding of their scientific meaning. These concepts were attributed to the gods and were thought of as spirit forces. In addition, they used the term *world* to mean the universe or "everything," and not necessarily the Earth.

Greek Theorists and Schools of Thought

Some of the Greek philosopher/scientists and schools were important to the development of modern chemistry and physics. The philosopher/scientists from about 700 B.C. to 400 B.C. are known as *presocratics.* Following are descriptions of the most influential.

The Ionian School

The Ionian School was located on a group of about 40 islands in the Ionian Sea, west of Greece and east of Italy. The Ionians invaded Greece from the north and settled in the islands west of the mainland, and also in southern Italy (Elna). The major Ionian cities were destroyed by the Persians in 530 B.C. The Persians brought with them new ideas that influenced the thinking of the inhabitants of the Greek mainland and islands. After Athens defeated the Persians, it became a prosperous city-state, which enabled citizens to develop new skills and inventions and concentrate on learning. Some historians consider Ionia the birthplace of Greek science and philosophy. The Ionian philosophers explained the world in physical terms, but most of their explanations were misconceptions. They were the first people to separate the Greek gods from material things in nature. They did not personalize the gods. They were wrong in many of their concepts, such as, space, astronomy, the origin of life, matter, motion, and the world. However, this group, which existed for several centuries, was the first to come up with the idea of the four *elements* (air, water, earth, and fire), which they believed combined in different proportions to form all the substances on the Earth. This misconception prevailed for another 2,000 years, into the Renaissance.

Thales of Miletus

Thales of Miletus (c.625–547 B.C.), one of the most famous Ionians of this period, established the Milesian School. Although Thales tried to integrate all things in nature conceptually (somewhat like Einstein's General Theory of Relativity or the General Unification Theory), his major misconception was that he believed everything was made of water, by which he meant that everything had a "mobile essence." He also believed that the Earth was a cylinder or a disk with the mountains, land, deserts, seas, and so on all floating on water. He also believed that loadstones (natural magnets) possessed souls because they could move small pieces of iron. Thales laid much of the conceptual groundwork that led to natural laws.

Pythagoras of Crotona

Pythagoras of Crotona (some historians give his birth/death dates as c.532–500 B.C., others as c.560–480 B.C.) was an Ionian who established a religious school in southern Italy that taught about the transmigration of souls from one body to another. Later, his school, which lasted for over 200 years, divided into two schools; one of religion, the other of science, where mathematics, astronomy, biology, and anatomy were studied. He is famous for work in geometry; in particular, his Pythagorean theorem for right triangles, which was partially worked out by Thales, and which is well known to students of plane geometry. The theory states that the square of the hypotenuse of a right angle is equal to the sum the squares of the other two sides

Pythagoras divided the universe into three components: (a) *Uranos,* or the Earth as a sphere; (b) *Cosmos,* which was the heavens surrounded by fixed stars, also in a sphere or shell; and (c) *Olympos,* which was the home of the gods. Another of his misconceptions was that everything is a perfect sphere or is arranged in concentric spheres, and that all motion of heavenly bodies must be uniform circles. This proved vexing when, many years later, it was determined that the planets, including the Earth, do not move in perfect circles, but rather in ellipses.

The Eleatic School

The Eleatic School of presocratic philosophers of the fifth and sixth centuries B.C. sought natural explanations rather than mythological legends to explain natural phenomena. They believed in movement (energy) and objects (matter) as separate units. They recognized that, in nature, there are basic units of things, but did not try to categorize or examine them. We will now consider several philosophers from this school.

Anaximander. Anaximander (c.611–547 B.C.), a famous student of Thales, was one of the early Eleatic philosophers who influenced later philosophers. He held several scientific misconceptions. One was that the Earth was unsupported and located at the center of the universe. He held two conflicting beliefs: First, he believed that all common forms of raw material or matter in the universe are eternal. Second, he also believed that over time, one common type of matter can change into another common form. He also believed that all life originated in the sea, which is a rather modern biological concept.

Parmenides. Parmenides (c.515–445 B.C.) stressed idealism and that numbers were paramount. Contrary to Aristotle, he believed that form (or function) was more important than matter (physical/chemical attributes). Idealism was the basis for Greek philosophy for several centuries.

Anaxagoras. Anaxagoras (c.488–428 B.C.), a follower of Thale's Miletian philosophy, held the misconception that all things are composed of air. He believed that the Earth was a cylinder, not a sphere, as did the Pythagoreans. He believed that as a first principle, the basis of all matter was the atmosphere. He also believed that all matter in the heavens is the same as on Earth, and that the raw material of matter is eternal, but changes into common forms of matter. One of his major misconceptions was that the sun was a large, very hot stone. He was the first to describe how the shadows on the Earth caused by the sun and moon are responsible for the solar and lunar eclipses. He did not think that matter or the heavens were divine.

Heraclitius. Heraclitius (c.550–475 B.C.) did not leave many records, but we do know that he believed in the four common forms of matter. He was called the "weeping" philosopher because of his pessimistic view that everything changes. He also said that the only permanent thing is change, making statements similar to: "Everything is in a state of flux," "Change is the only reality," and "There's nothing is and nothing was, but everything is becoming." He explained the processes of change in terms of retribution. As an example, he believed that all things are an exchange of fire, and fire is exchanged for all things. He believed that the major element was fire and that all fires are different and new, and that fire is the origin and image of all things. His main concerns were with the processes of change, not how the world was constructed. His idea of retribution was as commerce, which is expressed as *gold for merchandise, and merchandise for gold*; thus, an exchange. Another major misconception was that the sun of today will not be the sun of tomorrow because it is always replaced, it is never the same sun. He is remembered for his statement, "A man cannot step into the same river twice." Heraclitus asserted that reason was primary to all systems and philosophies.

Democritus. Democritus (c.460–370 B.C.) taught just the opposite. His major misconception was that he considered the universe as motionless and stable. He also believed that the Earth was a cylinder rather than a sphere. Democritus is often given credit for his insightful conception of atomistic theory, which stated that everything was made of "atoms," which are ultimate, homogeneous, indivisible, eternal bits of matter. But the credit should go to Leucippus (c.490–c.430 B.C.), his teacher, who first proposed the atomic theory in the fifth century B.C. Democritus just refined Leucippus's theory. This was centuries before the existence of atoms and elements was demonstrated. Their concept was that if you take some soil, divide it by half, then divide that by half, and continue dividing by halves until you have just one particle, it can no longer be divided. Several previous philosophers believed that you could continue dividing the soil, or anything else, forever.

The Atomists believed that life, including all animals and man, developed from some type of primitive slime or swamp, and that each person contained at least some of every kind of atom, some of which were always emitted, others always taken in by one's body. Their concepts of chemistry and physics were mechanistic; everything was predetermined or preordained—that is, everything is as it was meant to be.

Empedocles. Empedocles (c.490–430 B.C.) was influenced by the Pythagoreans. He believed that all reality was forever and all things were composed of a proportional combination of indestructible primordial matter: fire, air, earth, and water. One of his major misconceptions was that all chemical matter was combined or separated according to how much *love* and *strife* (opposing qualities) exist in nature; for instance, his four elements had four qualities he referred to as "cold," "hot," "wet," and "dry." Empedocles was one of the first to come up with the concept, later refined by Aristotle and others, that the four elements make up the universe. In addition, he believed that the atmosphere was not a void, but contained matter because of his belief that a finite universe (space) required matter in order to exist. He believed, as did most philosopher/scientists of this period, that infinity was reserved for the gods. Another misconception held by Empedocles was his concept that all living things first existed as body parts, for example, legs, arms, heads, organs, eyes, and so on, with all parts existing and moving around at the same time, and later "evolving" into total animals or humans. Even though Empedocles obviously held great misconceptions, he made one of the first contributions to modern physics: He stated that *quality* and *quantity* are separate entities and a change in an object's position changes the object itself (quantum theory and the Heisenberg uncertainty principle).

Zeno of Elea. Zeno of Elea (c.495–430 B.C.) was concerned with motion, space, and time—all of which are aspects of modern physics. But he held many beliefs about these concepts that have little relationship to modern science. He developed *Zeno's paradox*. In fact, he made a group of his paradoxes his philosophy. Zeno's paradox deals with a misconception of space/time. It is based on the concept that if you start to travel from point A to point B, you must first go *half* the distance toward the end (to the middle). Once at the middle between the two points, you must next go *half* of the remaining distance toward point B. Once you reach this middle point, you must again travel another *half* of the remaining distance, and so on, but you will never really get to point B because it is impossible by this reasoning. Here's the story related to it. It involved a race between Achilles (a mythological Greek hero) and a tortoise. Or as we know the story, it is a race between the tortoise and the hare. No matter, in both versions of the story if Achilles or the hare gives the tortoise a head start, as per Zeno's paradox, the fastest runner must always reach a point from which the slower runner began. Achilles (or the hare) can never overtake the slower tortoise. The tortoise will always be ahead, and neither will ever arrive at point B.

The three Greek philosophers who had the greatest influence on Western thinking are Socrates, Plato, and Aristotle. Let's take a brief look at some of their beliefs and misconceptions in science in general, and chemistry and physics in particular.

Socrates

Socrates (c.470–399 B.C.) taught a philosophy stating that humans should be taught truth, justice, and virtue as guides for living. He is considered a great moral teacher but not much of a scientist, but he did say "Knowledge is virtue," which to him meant "practical knowledge." Socrates developed a technique for teaching still in use today. It is called *dialectical reasoning,* which is based on a continual, in-depth form of questioning everything and examining contradictions. His questioning technique was a major scientific contribution still used to explore natural phenomena. Socrates left few written records, but was immortalized in Plato's writings and Aristotle's philosophy.

Plato

Plato (c.428–347 B.C.) was a student of Socrates and like his master was concerned with ethical motives. In 387 B.C. he established the first university in Europe on land belonging to a famous Greek named Academus. Therefore the university was called an *academy.* Plato's academy taught biology, astronomy, mathematics, as well as philosophy and politics. In addition to the academy, he contributed a technique of writing and thinking that involved a dialogue or conversation between two people as they tried to solve problems. This dialectical process became part of scientific inquiry. Plato believed mathematics was a form that could define the universe. The *Platonic Idea* was more concerned with the form of a concept than the matter itself. Unfortunately, much of the science taught in Plato's academy was based on misconceptions of reality. Scholars thought, talked, and argued about scientific things, but they still did not try things out to see what might happen.

One of the problems with the way the Greeks approached their understanding of the physical universe was their failure to analyze and pay attention to their own observations and make use of empirical experimentation. One example is their observation of the motion of the planets. They developed a system of unrealistic and complicated epicycles for planetary motion to conform to their mistaken concept of an Earth-centered solar system. Another example of how they ignored their own observations was by placing "beliefs" on a higher plane of reality than reality itself. They favored using abstract thought, metaphysics, mathematics, and logical deductions to experimentation. Aristotle, one of Plato's students, began to change the classical Greek/Roman philosophical approach to a more modern scientific approach.

Aristotle

Aristotle (c.384–322 B.C.) was a Greek philosopher and scientist who attended Plato's academy. His writings cover a whole range of knowledge. Aristotle was influenced by his father, who was a physician. In addition to medicine and mathematics, he had an interest in biology. He believed that each individual occurred in a fixed natural type (species). He believed that individuals have built-in growth and development patterns preordained by nature (genetics). One of Aristotle's contributions is the concept of *causality*. He thought that any event or thing had more than one explanation. His four *causes* are not exactly the same as our understanding of cause and effect, but they are instructive. Aristotle's four causes are as follows: (a) The "material cause" of which things are made; for example, earth, air, water, and fire (chemistry); (b) the "efficient cause"—the source of motion and energy that causes things to move; nothing moved unless it was being pushed by some outside force (energy, inertia, momentum, force); (c) the "formal cause," which relates to types or species of things (biology); and (d) the "final cause," which is the goal of full development of an individual, invention, structure, or belief (growth, technology, engineering, religion). He considered his causal concepts the key to organizing knowledge, and he tried to organize everything in nature by categories or classes.

Aristotle proposed many other misconceptions of science. He considered the universe to be spherical and finite. Another of his misconceptions that lasted hundreds of years is that the Earth is at the center of his universe. He believed that his four elements (earth, water, air, and fire) sought their own levels according to their relative heaviness (density). Another major scientific misconception Aristotle held was that if you had two bodies made out of the same substance, and of the same shape, but not the same size or weight, the larger and heavier body would fall to the ground first if both were dropped at the same time (inertia/mass, acceleration, force, gravity). This was a common misbelief based on "common sense" that was disproved some years later when Galileo used his "uncommon sense" to try it out. Aristotle also held many erroneous beliefs about biology, including "spontaneous generation" (see Chapter 3).

Aristotle made many contributions to the fields of philosophy, psychology, ethics, logic, and metaphysics. A major contribution was the concept that a person should base his or her conclusions on observations—not just on beliefs (deductive reasoning). One of the greatest compliments paid to Aristotle is attributed to Charles Darwin, a great admirer of Aristotle. Apparently, when Darwin was discussing the intellectual giants of history, he said of the scholars of his time that they "were mere schoolboys compared to old Aristotle."

There were a number of other philosopher/scientists after Aristotle who held a variety of misconceptions dealing with chemistry and physics, particularly the concepts of energy, force, and motion, as well as astronomy. Others made some contribution to science, but in general the few centuries before the birth of Christ were a low period for both Greek and Roman enlightenment. There were a few exceptions, however. Euclid (c.330–260 B.C.) gathered many mathematical proofs from various sources and published them as his *Elements of Geometry.* His work became invaluable for later scientists. Eratosthenes (c.276–194 B.C.) made accurate measurements of the size of the Earth that were important to navigation. Others who made progress in astronomy and physics were Apollonius (c.220 B.C.), and Hipparchus (c.190–120 B.C.).

The Concept of Motion

A major misconception held by most philosopher/scientists for over 1500 years, from about 500 B.C. to the time of Sir Isaac Newton (1642–1727), involved an aspect of physics—the concept of motion. The Greeks, and many others from later periods, considered motion to be caused by something either pushing or pulling an object, and when this "mystical force" stopped so did the object's movement, instantaneously. They did not believe in "action at a distance" that they couldn't see. Newton's three laws of motion and gravity (see the section on Isaac Newton in this chapter) finally explained motion, at least in a Euclidian context.

The last few centuries before the birth of Christ were also the beginning of the period of early alchemy in both medicine and chemistry. We will address alchemy in the next section, which explores the scientific developments and misconceptions of chemistry and physics in the Dark and Middle Ages of science.

THE DARK AND MIDDLE AGES OF SCIENCE

There was a gradual decline in science and intellectual inquiry from about 300 B.C. to about 1350 A.D. This is the period referred to as the Dark and Middle Ages of science. The year 1400 is considered to be the end of the Middle Ages and the beginning of the Renaissance. It should be remembered that these historical periods have no exact beginning or ending dates. Some historians suggest that the Dark Ages comprise the years between 400 and 1100, and that the period between 900 to 1200 was the development of Islamic learning. This is the period when science moved East as it declined in the West. This Age of Arabian Influence kept science alive through translations into Arabic of Greek and Roman literary works. Another milestone signalling the end of the Middle Ages of science was Sir Isaac Newton's (1642–1727) development of the three laws of motion and the theory of gravity.

The causes and events that led to the Dark Ages are as varied as the historians who proposed them. The Greeks developed unique ways of asking questions, and less personal, more objective ways of looking at things and thinking about what they saw. In the city-state of Athens, an environment of freedom flourished, as did democracy, political ethics, the arts, language, poetry, and just as important, ideas about nature. All of this had and still has a fundamental influence on Western civilization. The Peloponnesian War (431 to 404 B.C.) was the beginning of the end of the Greek city-states. There was a period of several centuries, referred to as the Hellenistic Age, which coincided with the conquest of Greece by the Roman Empire. Once the Greeks lost their independence, some scholars relocated in Asia Minor, the Aegean Islands, Egypt, and Italy. Even so, the Romans contributed to the demise of inquiring minds. They were more interested in rhetoric for its own sake, than in investigating nature. But they did make use of Greek knowledge to develop technologies, such as engineering and construction, which they used to build temples and aqueducts.

There are several other possible reasons for the decline of scientific inquiry besides the end of the Greek city-states and the fall of the Roman Empire. One possible reason is that during the Dark Ages no new, coherent, or scientific philosophical system developed as during the Hellenistic period. A major contributing factor was the rise of monastic religions, which were based on one god and epitomized by an individual human being. Because ancient people often thought of death, they wanted some assurance that they would not become sick and die. They wanted guidance in their lives as well as assurances, even to the extent of everlasting life. Therefore a variety of religions developed to offer answers and quell qualms. As mythologies developed they incorporated what some considered "the answers." These answers developed into proscribed activities, rituals, bodies of writings, prayers, magic, superstitions, sacrifices, as well the development of rules (religious laws) for disciplining people.

Christianity, Judaism, Islam, and some other religions contributed greatly to the dearth of science during the Dark and Middle Ages. Much of the opposition to science by religious people was from a misunderstanding of the purposes and practices of both religion and science. During this period many people held the misconception that religion could be based on doctrines that never change, and that the accepted knowledge about things as taught by a particular religion were true, even if the facts indicated otherwise. Science accepts facts, up to a point. When the data and evidence changes, or a better answer becomes available, science not only requires but demands change in keeping with the new knowledge. Religions have no such built-in self-correcting concept or procedure in their systems. Christianity, and several other religions, not only suppressed scientific inquiry and knowledge, but for over 1,500 years killed many great people because of their unacceptable beliefs. Curious men and women, using their inquiring minds, tried to find answers to age-old questions about themselves, their universe, and nature—often with great sacrifice.

Scientists of the Dark and Middle Ages

During the Hellenistic Age (fourth to first centuries B.C.) Greek became an international language, along with Greek culture, art, poetry, and science. Because of the decline of Greece as a nation of city-states, wealthy rulers of other countries, particularly Egypt and Macedonia, paid Greek scholars to bring Greek culture to their countries. Some of the philosopher/scientists of this period continued to explore astronomy, mathematics, medicine, physics and chemistry. A few examples follow.

Euclid (c.300 B.C.) moved from Greece to Egypt where he taught mathematics in the city of Alexandria. His move was typical of many of the most intelligent and knowledgeable Greeks who moved elsewhere during this period. Euclid's major contribution was his compilation of many geometric works by other mathematicians into an extensive work called the *Elements of Geometry.* By developing axioms for simple problems, he then systematically developed theorems from these axioms to prove more complex problems in geometry. Even though his work lasted over 2,000 years, there were misconceptions related to it. First, his geometry was related to only plane surfaces (two dimensions), which was corrected for three dimensions in later centuries. Two other real-world misconceptions are: (a) "the whole is greater than the parts," which in the physical world is not possible (you can't have something that is greater than its components); and (b) "if a point lies outside a straight line, then one, and only one, straight line can be drawn in the plane that passes through the point and which never meets the line." This supposedly described a line parallel to the straight line, but neglected to consider a third dimension.

Archimedes (c.287–212 B.C.) is the most well-known scientist of the Hellenistic period. He had many interests, including the study of plane geometry and cylinders, cones, and spheres. Archimedes spent much time figuring out the ratio of a circle's circumference to its diameter (pi or π) by "squaring the circle." His figure of 3.142 for π was very close. He developed war machines based on his knowledge of levers, pulleys, and optics to throw stones and concentrate the sun's rays to cause fires in the defense of his native Syracuse against the Romans. In a few years, Rome completed the invasion, and Archimedes was killed by a Roman soldier when he neglected to hear or respond to the soldier's order. He also contributed to the development of the concept of buoyancy.

The story of Archimedes and his discovery of specific gravity is well known, but it can be told again as an example of a revelation of a scientific concept to a person with an inquiring mind. A king named Hiero ordered a crown of solid gold to be made. He suspected the contractor cheated him by substituting some silver for the gold but could not determine the crown's exact content. He asked Archimedes to investigate the situation. While pondering a solution Archimedes lowered himself into a full bath; when the water ran over the sides, he noticed that the lower he went down into the tub, the more water spilled over. He immediately found the solution to the problem. As it is reported, he ran down the street shouting "Eureka! Eureka!" ("I have found it! I have found it!") He then made a mass of silver and one of gold, each the same weight as the crown. Then he filled a pan to the brim with water and lowered the lump of silver into the pan, collecting the spilled water, and did the same for the lump of gold. He then measured the volume of the overflow water for each. The smaller amount of spilled water was caused by the gold, since it is more dense and thus less in volume than the identical weight of the silver lump. He repeated the process with the crown, and found that more water ran over the top for the crown than for the equal weight of gold, and less water was displaced by the crown than for the silver lump—thus the crown was not pure gold. One version of the story relates that the craftsman of the crown was executed. This experiment of Archimedes demonstrated the relative densities of gold, silver, and a mixture of the two.

It is difficult to identify any major misconceptions held by Archimedes. There is some evidence that he held several Aristotelian views of astronomy, including the geocentric concept of the universe surrounded by celestial spheres. He did not investigate the chemistry of matter and accepted the misconception of the four types of elemental chemical matter (air, water, earth, and fire). He was, however, familiar with the three states of matter: solids, liquids, and gases. Much of his work was accepted and did not need correcting or refinement until the scientific revolution of the seventeenth century when Sir Isaac Newton developed his laws of motion.

ALCHEMY

Alchemy developed slowly in the waning years of the Hellenistic period and continued to develop as a pseudoscience and a philosophy for the next 1,500 years. The study of chemical changes (alchemy) started more or less simultaneously in Egypt, China, India, Babylonia, and Islamic countries, including Syria and Persia. Early on people developed the ability to extract copper, gold, silver, lead, mercury, and arsenic from their ores, which soon became the basis for alchemy.

Historically, there are two forms of alchemy—the *exoteric* and the *esoteric*. Exoteric alchemy is concerned with using chemicals to form the philosophers' stone, which could then be used in the transmutation of base metals into gold. The esoteric was more metaphysical, based on superstitions and astrology. The stone, also known as the *elixir vitae* (elixir of life), was thought to be the fifth element that could be added to base metals to produce tinctures (alloys) that could also be used as the ideal cure-all (sovereign remedy) for all diseases and provide long and eternal life. The philosophers' stone was also known as the "medicine of the metals." It was mistakenly believed that it not only had healing power but could change base metals into gold and silver—if it could ever be found. In time, alchemy became more of a religion, mixing the chemical aspects with the theological, the mystical, and the astrological beliefs of its practitioners.

Alchemy was based on the idea that matter has a soul that can be moved from one element to another, if only the stone that would accomplish this process could be found. Much time, effort, and money by the rich was expended over the years to perfect this process. It was believed that if fire and water were purified, the stone would be produced and thus be able to change lead into gold. But if the mixture was somewhat impure, then silver would be produced. Although many unscrupulous people faked the production of both the stone and gold, it also led to a better understanding of the nature of matter and how different substances react with each other. In other words, alchemy might be referred to as the antecedent to modern chemistry.

Alchemists learned how to produce acids, how to use heat to control chemical reactions, and how to distill and purify certain chemicals. They developed the process of distillation to a fine art. Many alchemy laboratories of the Dark and Middle Ages were well equipped. Several pieces of alchemy equipment and techniques, although refined, are still used in modern chemistry labs.

The Dark Ages of science are epitomized by the rise and popularity of alchemy. Numerous false beliefs and misconceptions were advanced by many different people. Let's examine some of the misconceptions that abounded during this period.

We have previously mentioned Aristotle's achievements, but let's look as some of his misconceptions related to alchemy. He thought that all matter had "form" in four primary qualities expressed in the following arrangements: (a) hot and dry are assigned to the element *fire*; (b) hot and fluid are assigned to *air*; (c) cold and fluid are assigned to *water*; and (d) cold and dry are assigned to *earth*. He held the misconception that all elements are composed of these four "forms." He also mistakenly believed that the only factor that makes substances different from each other is the proportion of one form (substance) to another. Other major misconceptions were that if a substance was able to burn, it contained its own "fire"; if it was a thin liquid, it contained its own "water"; if it was gaseous, it contained its own "air"; and if it was stone it contained its own "earth." Aristotle provided the basis for the concept of transmutation in alchemy with his belief that one substance can be changed into another by changing the proportions of its elemental forms. Later, alchemists believed that a spirit of some type, through astrology, could be used to produce the philosophers' stone, if only they could discover how the process worked.

There are several other early scholars of alchemy. Zosimus of Egypt, who has already been mentioned, summarized ancient alchemy and experimented with mercury, arsenic sulfide, copper arsenide, white lead (litharge plus vinegar), and natron (sodium carbonate). Stephanos of Alexandria saw alchemy as a way of mystical thinking, not necessarily as being involved with chemicals. Hierotheos and Heliodorus wrote about alchemy as the mystery of transmutation. In China, Dzou Yen claimed to be able to produce gold by alchemy. He failed, and in 144 B.C. alchemy was banned in China.

Several Arab scholars, familiar with Greek and Egyptian science, compiled much information on alchemy and other sciences. Jabir ibn Hayyan compiled a book of all the knowledge of alchemy and other subjects in the eighth century. Another Muslim scholar, al-Kindi, translated some Greek works into Arabic. He was interested in astronomy as well as astrology. He knew about alchemy, but was a skeptic. One of the best-known Islamic alchemists was Khalid ibn Yazid (c.660–704), who studied in Alexandria. He worked with associates in a well-equipped laboratory to produce gold by transmutation. He ordered the execution of an alchemist who claimed that he had successfully produced gold, but had not.

The period between the eighth and eleventh centuries were known as the Golden Age of Science in Arabia. During the twelfth and thirteenth centuries Arab scientists translated Greek works into Arabic and other languages, marched west and progressively invaded other countries. They brought along their science, which was based on translations of the original Greek writings, and "imported" alchemy into Spain and other European countries. This was also the period when alchemists were improving their laboratory vessels and distillation techniques. The arts of metallurgy, glass making, cloth dying, and the discovery of several elements and compounds such as sodium hydroxide, ammonium chloride, as well as various arsenic, mercury, and antimony compounds also occurred during this time. It was about this time that the Muslim religion began to spread throughout the Arab world.

There is some dispute among historians as to the degree that science was transmitted from Islamic to Western countries. It is assumed that alchemy was not known in medieval Europe until it was introduced by Eastern cultures, but this is not proven. Arab texts were not translated into Latin for use by Europeans until 1144 when Robert of Chester undertook the task. Robert's translations of Muslim alchemy were mostly from Khalid ibn Yazad's works.

Several people in the late Middle Ages tried to organize the multitude of chemicals and processes used by alchemists into a reasonable system. Paracelsus stated that only seven distinct processes were required to organize alchemy. One system devised independently by George Ripley was based on the 12 zodiac signs. The alchemical processes related to the signs of the zodiac are as follows:

1. Calcination > Aries, the Ram
2. Congelation > Taurus, the Bull
3. Fixation > Gemini, the Twins
4. Solution > Cancer, the Crab
5. Digestion > Leo, the Lion
6. Distillation > Virgo, the Virgin
7. Sublimation > Libra, the Scales
8. Separation > Scorpio, the Scorpion
9. Creation > Sagittarius, the Archer
10. Fermentation > Capricornis, the Goat
11. Multiplication > Aquarius, Water Carrier
12. Projection > Pisces, the Fish

Let us now briefly consider several important figures in Western alchemy.

Albertus Magnus (c.1200–1280) tried to combine Aristotelianism with Christianity. His contribution to chemistry was the production of arsenic free of impurities.

Roger Bacon (c.1220–c.1292), an English scholar/philosopher, made accurate observations of natural processes. He was interested in alchemy, and his misconceptions are related to his teaching of magic. Bacon was ahead of his time, and some of his contemporaries accused him of being a sorcerer and necromancer because of his experimentations. He and his cohorts constructed a head out of dead human-head pieces, which they expected would speak to them, with the Devil's help. As the story goes, Bacon stayed up night and day waiting for the head to speak. Finally, he assigned a servant to observe the head. When the head supposedly did speak, the servant was afraid to awaken Bacon because all it said was "Time is," which he thought was not an important-enough message for which to awaken his master. Bacon was a great experimenter and was thought to have possibly invented gunpowder (which was already known in China), as well as the compound microscope, which he supposedly used to observe cells of living organisms. These claims were never proven. It might be mentioned that alchemists' experimentation was not as we think of it today. We use experimentation to infer or confer general scientific laws about specific examples. Alchemists used experimentation in the Aristotelian sense, to draw conclusions from principles that were already established through observation, thought, or philosophy (deduction).

Saint Thomas Aquinas (1225–1274) was a pupil of Albertus Magnus. He followed his mentor's lead in facilitating many of Aristotle's concepts with Christianity. Like Aristotle, St. Thomas saw humans as a total of body and soul. Both held the misconception that resurrection of the body was not only a philosophical reality, but a real possibility. He accepted Aristotle's concept of intellectual knowledge as derived from sensory perceptions or abstractions. St. Thomas believed that the existence of God can be proven by sensory data and by accepting revelation as the basis of human knowledge. One of his misconceptions was the belief that if some things related to God have been revealed as true, then it is reasonable to accept all unknown mysteries of the universe as being related to God, and thus true.

William of Occam (c.1285–1349) was an English philosopher/theologian who accepted Aristotle's idea that science can be demonstrated using basic principles. He rejected the Thomistic view of science, which was based on the rational acceptance of God's existence, the divine, and the immortality of the soul. Occam is best known today in the areas of physics and chemistry for his dictum called *Occam's Razor,* which states that if there are many reasons proposed to explain something, they should not be used if something simpler suffices. Or, "do not do with more what can be done with fewer." In today's vernacular we call it KISS (keep it simple, stupid). Occam preceded Isaac Newton with the concept that planets (or any body) in motion do not need physical contact with a mover (push or pull) to keep them moving. He also rejected the commonly held misconception that a group of angels was responsible for the movements of the planets.

An important scientific development that no doubt started in Eastern countries was the use of chemicals to treat illnesses. There are two important figures in the late Dark and Middle Ages who contributed to this aspect of alchemy. They have also been referred to as Renaissance men because of their wide ranges of interest and their works, which some say marks the beginning of the Renaissance, also known as the Age of Enlightenment.

Leonardo da Vinci (1452–1519) was a great artist who was also a man of science. He investigated the natural (botany, zoology, anatomy, etc.) and the physical sciences (mathematics, mechanics, engineering, etc.). He was one of the first to try to combine all science into one giant theory. One of his goals was to develop a unified theory of the universe and then illustrate it in a large book. He became so busy, he never finished all his projects, including his "unified theory of everything," which scientists are still pursuing to this day. Although his drawings of the human body were remarkable, many of his concepts of chemistry, medicine, and anatomy were primitive. Without knowing the principles of mechanics or physics, he made remarkable use of these sciences in engineering, architecture, and construction of large public works.

Theophrastus Bombastus von Hohenheim, better known as Paracelsus (1493–1541) was one of the last great alchemists (see the section on Paracelsus in Chapter 2). Alchemy continued on the fringe of science for several more centuries, but Paracelsus was one of its masters. For his day he was an accomplished alchemist and physician. Even so, he believed in astrology and some of the mystical aspects of science. He developed many diagnostic techniques and treatments of various diseases, not all of which were useful, and some of which were even harmful. One of his major misconceptions was that mercury and some forms of lead could cure several diseases. He and many other physicians of his day often poisoned their patients or bled them to death. He might be thought of as the first consumer advocate of medicine, because he attacked the ethics of so many of his colleagues. He is best remembered for his contributions to chemical physiology and chemotherapy, and might also be called the father of modern pharmacology.

By the end of the Dark and Middle Ages alchemists became very diverse in their concepts. Many alchemists were learned philosophers and visionaries. Some believed in the occult, magic, and superstitions. Still others became practical alchemists who were interested in understanding chemistry, pharmacy, and nature in general. These alchemists led the way to modern chemistry and medicine.

During the Dark and Middle Ages when alchemy was the main focus of investigation, philosopher/scientists interested in the natural sciences started to conduct chemical experiments to learn more about the makeup of matter and how different substances react with each other, rather than to just search for the philosophers' stone or elixir to produce gold and medicines. The first textbook written about chemistry, *Alchemia,* by Andreas Libavius was published in 1606. This was the first record of chemists trying to change alchemy from a philosophy into a science. It instructed one on how to prepare reagents (chemicals that can react with other chemicals in order to identify and measure them), and how to extract pure substances from mixtures (distillation, etc).

Before chemistry and physics could progress beyond this period, many corrections were required, not only of factual misconceptions, but also in the way people thought about nature. It was years before progress could be made.

THE RENAISSANCE OF CHEMISTRY AND PHYSICS

This period of enlightenment that coincides with the rise of humanism, which was an attempt to rationalize Christianity by the revival of ancient learning—particularly Aristotle's works, which over the next few hundred years led to the rise of "the experimental way." Another feature of this period is the changing relationship between technology and science. Up to the time of the late Renaissance, the science that developed was accomplished according to the technologies known to the respective age. The sixteenth and seventeenth centuries comprise a period whereby science began using technology as a tool, rather than the other way around.

Bubonic plague or the "Black Death" devastated Europe in the middle of the 1300s, killing half the population. This disaster forced scientists to learn more about nature, medicine, chemicals, and how things worked based on observable facts, not philosophical beliefs. One of the great inventions that enabled the spread of knowledge was the development of moveable type (the printing press) in 1440 by Gutenberg.

This invention led to an explosion in the publication of books and increased the ability of chemists and physicists to better communicate their ideas. Exploration by land and sea was increasing, which required better knowledge of the Earth, the heavens, and the oceans. This led to inventions that enabled scientists to more precisely measure and analyze natural phenomena (the telescope and microscope) and to improve navigation (sextant, chronometer). Increased travel spread diseases, as well as knowledge. In keeping with the concept of change, Martin Luther's 95 theses introduced the Protestant Reformation in the year 1517. Protestantism was somewhat more open to scientific inquiry than was the Catholic Church.

Francis Bacon

Francis Bacon (1561–1626) is often credited as the founder of the scientific method. The steps he took toward scientific inquiry were: (a) classify all science; (b) use inductive logic; (c) gather experimental and empirical facts; (d) use examples or analogies to explain the phenomena; (e) make some generalization based on known natural history; and (f) correct errors by conducting future experiments and collecting new data.

Preceding Bacon were several scientist/philosophers who believed in the occult, the mystical, and what was known as natural magic, which was the precursor of natural philosophy or the natural sciences. It was thought that magic had two forms: the one filled with foul spirits, wicked curiosity, and chicanery, known as sorcery; the other was considered the magic of nature. Natural magic included the study of magnetism, mechanical arts, optics, the study of stones, herbs, mathematics, and astrology. The proponents of natural magic were said to practice *archemastry*. Archeus is a term used by Paracelsus to represent the spiritual force of a seed or growth in a body (like a miniature alchemist). Archemasting is the mastery and use of the concept of archei. They also considered science an art. John Dee (1527–1606), who was considered an archemaster, wrote the first English translation of Euclid's mathematics. He also preceded Francis Bacon in recognizing the importance of experimentation as well as observation before making conclusions about nature. This was when natural magic and true experimental science parted company.

Francis Bacon was both an English statesman, holding several offices in his lifetime, and a philosopher with a belief that people are the interpreters of nature, and that truth and knowledge are derived by experience—not authority. He was not considered a great scientist, even though his philosophy influenced many generations of scientists. One of his books, *Novum Organum*, proposed an inductive philosophy designed to take the place of Aristotelian deductive philosophy. Bacon's scientific method made use of analogies to infer specifics from the properties of a larger group of data. One of his most important concepts was the idea of self-correction. He believed that later experience and experimentation would correct errors in current scientific knowledge as more accurate data became available. He is rightfully called the father of the modern scientific method.

Bacon's Contemporaries

Several other sixteenth-century scientists, not all known specifically for their contributions to chemistry or physics, influenced several scientists in the later Renaissance period. These scientists helped set the stage for later scientific developments in our own times.

Leonard Digges (c.1520–1559) was an English mathematician and surveyor. He is often credited as the inventor of the telescope. William Gilbert (1544–1603) was an English physician and physicist who made some of the first scientific studies of magnetism, including the Earth's magnetic field. He discovered that you could make a magnet by stroking a piece of iron with a lodestone or while hammering the end of an iron bar as it pointed in alignment with the Earth's magnetic field. One of Gilbert's misconceptions was that stars revolved around some central point; for example, the Earth. Tycho Brahe (1546–1601) and Johannes Kepler (1571–1630) were both Copernicans who added much to our understanding of the physics of the universe. Tycho used the telescope to make very accurate records of the stars and their positions. Kepler, who had poor eyesight, did not use the telescope extensively, but using Tycho's data he produced brilliant mathematical models to explain planetary motion that are still valid today (see Chapter 5 for more details).

A misconception still held by many scientists of this time was that there were multiple crystal celestial spheres that contained the moon, planets, and stars, and that their position at the time of your birth determined events in your life. Even Galileo was in great demand for his skill in developing astrological prognostications for the rich and famous. Because of the invention and use of the telescope and new mathematical principles, many scientists began to consider Aristotle's concept of multiple crystal celestial spheres invalid. They still held the misconception that there was at least one final, outermost containing sphere that represented the edge of the universe. Advances in astronomy and mathematics during the Renaissance provided much of the background knowledge for the development of physics and chemistry.

Galileo Galilei (1564–1642) followed the footsteps of Copernicus, Tycho, and Kepler in advancing knowledge during the Renaissance. He is better known for his work in astronomy, and we will return to his work in chapter 5. Even so, he was more interested in mathematics and mechanical systems than he was in astronomy. His contributions to physics are many. He is credited with discovering the laws of falling bodies and applying these laws to the parabolic path and motion of artillery shells. He also investigated hydrostatics, the motion of pendulums, and the mathematics of mechanics, which predates Newton's work.

His experiment of dropping two balls of different weights from the leaning tower of Pisa is pure legend. It is more likely that he actually rolled the balls down a low incline plane and timed their descent (which was slower than if he would have dropped them from the tower) to the bottom of the plane. Because accurate clocks had not yet been invented he used his heart beat as a stopwatch. He disputed much of Aristotle's work relating to both physics and astronomy, which made him unpopular with some of the other scientists of the day. His concept of motion provides an example. Most scientists believed in the Aristotelian idea that if an object moved, something had to either push or pull it to make it do so. When the push/pull stopped so did the movement—instantly. Galileo's theory of inertia states that if the body is moving, it will keep moving, or change direction, in relation to the force applied. These concepts were refined by Newton's laws of motion. Galileo's main contribution is his belief that scientific investigation must be open and not under the authoritarian control of politics or religion. One of Galileo's major misconceptions was his belief that light travelled at infinite speed and moved instantaneously from its source to the object receiving the light. He tried to devise a crude method of determining the speed of light, but failed.

René Descartes (1596–1650) is better known for his mathematics (analytic geometry) and philosophy ("I think, therefore I am") than he is for his physics. He held several misconceptions. One was related to his Cartesian circle, which states that a person cannot prove that God exists unless that person can be certain of his or her own thinking and reasoning that God exists (tautology or circular reasoning). He believed that one's body can be divided into an infinite number of parts, whereas the mind is whole and indivisible. He also believed that the human mind moved little parts in the brain that caused other parts of the body to move, and that the motion in the brain caused emotions and physical sensations. He also believed that physical sensations like smell, sight, sound, and so on only existed in thought. Another of his major misconceptions was that there was no such thing as a vacuum, because by definition, three dimensions cannot exist in an empty space. He considered gravity to be a swirling vortex that caused things to descend toward the Earth. He postulated that all phenomena such as magnetism, movement and collision of bodies, gravity, and so forth could be explained by what was then known as mechanistic physics. One of his contributions was his concept of inertia, even though he related inertia to the goodness of God.

One of Descartes's major misconceptions in the field of physics was related to the Aristotelian impetus theory of motion, which said that after an object began to move, air just kept pushing it along. This theory ignored the concept of universal inertia. Descartes believed that any object in motion had to be affected by a medium (air). The medium's impact had to keep in contact with the object to keep it moving. If the medium gradually decreased its impact on the object, the object would slow down. Descartes was ignorant of the fact that motion (inertia and momentum) exists in airless space; for example, a vacuum. (See the section on Isaac Newton in this chapter for the current explanation of motion.)

Robert Boyle (1627–1691) is best known for *Boyle's Law,* which states the inverse proportional relationship for the pressure and volume of gases. Robert Hooke (1635–1703) helped with Boyle's studies of gases by experimenting with air pumps. Together they studied the process of combustion. A misconception held by several scientists of the period, including Boyle, was the phlogiston theory of combustion proposed by Johann Joachim Becher and Georg Ernst Stahl. In essence, this theory states that all substances that burn contain phlogiston, and when combustion takes place there is a loss of phlogiston. Stahl's misconception was that fire was a material substance.

Boyle published *The Sceptical Chymist* in 1661, which brought to an end the Age of Alchemy. Even so, many still held the belief that there were basically four elements and three processes that could explain all chemistry.

Another of Boyle's beliefs was that "primary" corpuscular matter formed all other bodies. His belief was that "primary corpuscular matter" was the basic "stuff" of matter, similar to atoms, but also just one of the three or four "elemental forms of matter." To explain the different characteristics of substances, he postulated that each specific substance had a different amount of primary corpuscular matter. For all his misconceptions and his misguided belief that air was a single "elemental" primary corpuscular substance as well as believing the universe to be a big piece of clockwork, he was one of the most knowledgeable chemists of his time. Boyle founded the English Royal Society (of science) in 1662.

Isaac Newton

Isaac Newton (1642–1727) might not have developed his famous laws of motion if it were not for the plague in England in 1665. During this period Cambridge University was closed so he went back to his home in Woolsthrope where he spent much time working on his mathematics of mechanics, which germinated his ideas on space, time, and motion. He was influenced by Nicolas Copernicus, Johannes Kepler, and Galileo, whose works will be discussed in detail in Chapter 5. He used their ideas to formulate a mechanistic universe, somewhat similar to Boyle's, which influenced thinking up to the time of Planck and Einstein. He was familiar with the philosophy of Descartes, which affected his concepts of science, religion, and mysticism. It might be said that Newton's contributions to science were more revolutionary during his lifetime than were Einstein's.

In order to derive his laws of motion, Newton was required to come to grips with both the old and new universal concepts of time, space, speed, velocity, acceleration, force, mass, inertia, and gravity. His theories of classical mechanics were published in 1687 in *Philosophiae Naturalis Principia Mathematica,* which is still referred to today.

In essence, his laws of motion are: The law of inertia, which states that a body at rest tends to remain at rest, and if in motion, continues to move in a straight line in the direction of the force that caused the motion. This means that a state of rest is included in the concept of the state of motion. The inertia of a mass has a tendency to resist acceleration, or to resist a change in its constant motion (velocity).

The second law deals with acceleration of a body once in motion. It states the mathematical relationship between the force applied to a body and its acceleration. In other words, the body is accelerated in the direction of, and to the extent of the magnitude of the force ($F = ma,$ where F = force, m = mass, and a = acceleration). This means if you apply a force to a mass, the mass will change its direction to correspond to the direction of the force, and it will change its rate of acceleration according to the magnitude of the force applied to it. This equation can also be used to determine the mass of a body if its acceleration and the force applied to it are known, or to determine the

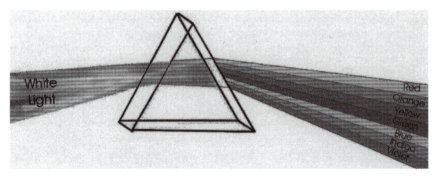

Figure 4.1 In the year 1666, Sir Isaac Newton experimented with prisms to demonstrate that white light (sunlight) is made up of different wave lengths represented by the colors of the rainbow. His work in optics led to his theory of color.

acceleration of a body if its mass and the force applied to it are known. It should be noted that the nature, origin, or type of force is immaterial; only the magnitude and direction of the force on the mass are important, which makes it a vector quality. This second law influenced physics until the theories of quantum mechanics and relativity of the late nineteenth and early twentieth centuries.

The third law is also simple, but is often misunderstood. In essence, it states that forces are equal but in opposition (as pairs) to each other. The statement, "for every action there is an equal and opposite reaction" is how most of us think of this. These laws led to the important concept of *conservation of momentum,* which states that two or more bodies stay at rest, if not in motion (inertia at rest), and they remain at rest until some force causes the mass to move and change its direction in the same direction as the force. Thus, the product of the mass and velocity of a body is called its momentum. For example, if one object that has a velocity (is moving at a certain speed in a particular direction) strikes a second object, part of the first object's velocity or energy will be transferred to the second object. No momentum is lost. The conservation of momentum is important when studying the interactions of two or more bodies, whether they are stars, subatomic particles, billiard balls, or a truck colliding with a car.

In addition to his three laws of motion, Newton developed the mathematics of calculus, which helped him describe his laws of motion. Calculus was also independently derived by Gottfried Wilhelm Leibniz, which caused Newton to delay the publication of his work. But Newton was given credit for its development because he actually developed his calculus before Leibniz. Newton also studied optics and explained the nature of the white light spectrum (see Figure 4.1). His theory of universal gravitation was a major contribution to physics. Newton's experiments with light led to his *corpuscular theory*, which was not accepted by most physicists who believed that light was composed just of waves—not particles.

The laws of motion led to Newton's concept of gravity, which, in simple form, states that if you jump off the roof of a house the force of gravity of the Earth (a very large mass) not only pulls you to the ground, but your body's gravity (a very small mass compared to the Earth's) pulls the Earth to you. This leaves little doubt as to whose mass is most important to the concept of gravity. Newton also calculated the effects of distance on the force of gravity. The pull (force) of gravity is *directly* proportional to the product of the masses of two bodies ($m_1 \times m_2$) and *inversely* proportional to the square of the distance between the centers of the two bodies. Gravity is universal, and it has explained much about the motion of the stars, comets, planets, dust, and so on in our universe. But gravity does not explain the motions of subatomic particles. Because the gravitational attraction between the nuclei and electrons of atoms and all the subatomic particles is so small compared to the other forces acting on them, gravity is ignored when studying subatomic particles. As we will see, other theories were developed to explain the relationships of the very small bits of subatomic matter to each other.

Newton held several misconceptions. Some were based on a lack of data and information. A major misconception was his concept of space and time as separate entities. He thought that both space and time were absolute and unrelated, as proposed by Euclid and others. It was not until the quantum and relativity theories were developed that this misconception was corrected. Newton was the first to provide a theory for the propagation of light, even though his corpuscular theory of light was discarded in favor of a wave theory that includes photons, which more accurately describe the behavior of light.

Newton was influenced by the ancient Greek philosophers whom he studied at Cambridge. At one time he held the misconception that there existed a subtle spirit within bodies that caused particles of these bodies to attract each other. Some authors claim that he believed (but only tentatively) that some form of ether was required in space to explain the transmission of things such as gravity, magnetism, light, and heat across space. The ether (or æther) is an ancient concept of an infinite, pervasive, elastic massless substance that exists in space and allows electromagnetic waves to travel from one place to another.

Another misconception was his belief, according to his corpuscular theory of light, that light travels faster in a dense medium (water, glass), than in a lighter medium (air, vacuum). A major limitation of Newton's laws of motion is that he based them on the Euclidean concepts of plane geometry. In other words, his laws only apply to two-dimensional plane surfaces and in straight lines. Even so, Newton's concepts of gravity, inertia, momentum, and angular momentum, as incorporated in his laws of motion, could account for the motion of planets around the sun. Newton was interested in alchemy, the occult, mysticism, and he practiced these all his life.

Newton's Contemporaries

Several of Newton's contemporary physical scientists disagreed with his theories. They also held several misconceptions. A listing of a few of these scientists follows.

Robert Hooke (1635–1703) was an inventive scientist. He developed a clock drive for telescopes, the iris diaphragm to sharpen the images of optical devices now used for cameras, a spring-driven clock, and a telescopic sight. A major invention of his was the vacuum pump used by Boyle and other chemists who studied gases. His two main misconceptions were his belief that the vibrations of light waves (frequencies) were at right angles to the direction of the propagation of the light. Additionally, he believed that the motion of the planets was a problem of mechanics; therefore he came up with an incorrect theory of gravity to explain their movement.

Christian Huygens (1629–1695) and Isaac Newton were not the best of friends. Huygens criticized Newton's work, which caused Newton to delay publishing his famous *Principia.* Huygens published papers on centrifugal force and concepts of motion that were later developed as components of Newton's three laws of motion. Huygens developed the pendulum clock and improved telescopes, which he used to examine the planets. His theory of the wave motion of light stated that light travels slower in a dense medium than in a less dense medium. This was correct and made Newton's corpuscular theory incorrect, and thus outdated. But Huygens held the misconception that for light to be waves it was necessary for some medium to be available in space to conduct the light, thus was born his concept of the "æther." Neither Huygens nor Newton believed that light could travel as waves in a vacuum. This is why Newton came up with the corpuscular theory of light, and Huygens came up with the concept of an ether existing in the vacuum of space. The controversy went on for some years. Neither concept is correct, independently. Modern physics considers light to be both quantum particles (photons) and waves with momentum (electromagnetic radiation), depending on the math and situation in which light is involved. Huygens major misconception dealt with his concept of gravity. He felt that because there was no visible mechanism connecting any two bodies, there could be no force exerted between them, as proposed by Newton.

Ole Christensen Roemer (1644–1710) measured the speed of light by comparing the time interval between one eclipse to the next of one of Jupiter's moons. His figure of 186,000 miles per second was very accurate. Today's figure is 2.9979×10^{10} cm/sec, which is expressed as c, which stands for the universal mathematical constant of the speed of light, as in Einstein's $E = mc^2$.

As a follow up to Newton's work, concepts of energy and force were proposed by a number of physicists of this period. Leibniz, Huygens, Euler, Laplace, and Lagrange all understood the relationships between Newton's laws of motion and length, velocity, acceleration, force, mass, and the conservation of momentum that led to the concept of energy. Their discoveries led to the idea that energy is equivalent to the physical concept of work, which gave rise to modern mechanics. Others still held the misconception that a term they called *action* is the product of the distance a body travels multiplied by its speed (instead of momentum). This misconception defeats nature's principle of economy of effort and time, which when combined state that chemical and physical events will always take the easiest and shortest path, and therefore will be completed in the least time possible.

Historical Events Affecting Chemistry and Physics During the Renaissance

The Chinese invented gunpowder around the eleventh century for use in firecrackers. They also used gunpowder to make small rockets, which they sent scooting across the ground to scare the horses of the attacking enemy. This was the first account of explosives used in warfare. It was not until about 1450 or 1500 that gunpowder was used in artillery and portable firearms in European warfare, and it was much later when the chemical reaction involved for exploding gunpowder was understood.

The Chinese are also credited with discovering the magnetic properties of lodestones. Legends recount that Marco Polo returned to the West with the Eastern concept of the lodestone, or a piece of iron magnetized by a lodestone, which when floated on a piece of wood in a bowl of water would point north and act as a compass. He is also credited with bringing back developments such as the mining and burning of coal, large-scale iron production, paper-making, moveable type, as well as other technological advancements long before they were developed in the West. Because of the antitechnology and antiscience philosophies and religions of the West, Marco Polo's discoveries were not accepted for several centuries. During the Middle Ages and the Renaissance, the West benefited more from knowledge coming from China, India, and Arab countries than the East benefited from knowledge travelling out of the West.

THE GOLDEN YEARS OF CHEMISTRY AND PHYSICS

The golden years extended the scientific discoveries of the preceding three or four centuries. This period might also be thought of as post-Newtonian or pre-Einsteinian science. There were several important developments, both in the sciences and society, that made the 1700s, 1800s, and early 1900s unique in the advancement of physics and chemistry.

First, there was a change in the dogmatic adherence to religious philosophies and social/political customs. This change made the exploration of science a viable profession. The quest for knowledge became acceptable. The second major change was the invention or improvement of a great many instruments that made it possible to more accurately measure scientific phenomena, such as the microscope, telescope, spectroscope, and so forth. Third, rational systems of measurement were developed, mainly for physics, but were useful in all the sciences. An excellent example is the metric system. Finally, scientists began to look for and see patterns and relationships between events and objects in nature at both the micro and macro levels, and they began to theorize and generalize from these observations. This enabled them to clean up and clarify many of the misconceptions and unsubstantiated beliefs proposed over the past 2,000 years by philosopher/scientists. Several examples of these misplaced beliefs are a geocentric, flat-Earth solar system surrounded by numerous celestial spheres, alchemy and the quest for the philosophers' stone, incorrect concepts of physical/spiritual actions, force, motion, and matter, and incorrect concepts of anatomy, physiology, human illnesses, and so on. And fourth, the invention of calculus.

The increased understanding of the nature of physics and chemistry led to the explosion of technology, which continues today. Modern technology, which led to the industrial and microelectronic revolutions, is based on scientific findings made during the golden years of physics and chemistry.

The scientific developments of this period made the connections between chemistry and physics even less clear. For instance, the interactions of atoms and molecules of chemicals are based on concepts of energy, which is usually considered an area of physics (as is the structure of the atom's nucleus). Science has a somewhat arbitrarily organized system of natural phenomena. People have organized natural structures, causes, and events into general categories and an increasing number of specialized fields, even though nature makes no such distinctions.

The next section presents developments and misconceptions in physics and chemistry that cover the two and half centuries of the golden years. As an example of the developments in physics over this period, we will consider electricity and magnetism and the importance of their relationship to later areas of science. For chemistry, we will follow developments and misconceptions of the structure of the atom, the identification of elements, and the chemical interactions of electrons in elements and molecules.

Electricity and Magnetism

There is some evidence that about 5,000 years ago the Chinese used lodestones and magnets made from pieces of iron that were rubbed with a lodestone as compasses. In 1,600 B.C., the ancient Greek, Thales of Miletus, knew that, if amber was rubbed with a silk cloth, it would attract bits of straw and lightweight particles of matter. Ancient people were also familiar with the deadly shock of lightning and the milder shock of electric eels. But there was no understanding of magnetism or electricity, and the phenomena of shock or attraction they observed between objects was attributed to some unseen spirit.

These two attractive forces created a great deal of interest among scientists and the general public in the seventeenth and eighteenth centuries. Parlor shows were conducted in both social situations and laboratories to demonstrate the new phenomenon of electricity at rest, known as static electricity. It was demonstrated that if one rubbed a glass rod with silk and then brought the rod close to a small ball of "pith" or other light, "pulpy" material, the ball would be repelled by the rod. If this was done with two balls that were separated from each other, that were then brought together, they would also repel each other. This indicated that some repellent "charge" was produced. If one rubbed the rod with wool or fur, a charge was also placed on the balls. But when one of the balls from the first experiment was brought close to a ball from the second, they would attract each other, demonstrating that they must be possessed with different, or attracting charges. Thus you could cause different types of charges by rubbing different types of substances together. It was later discovered that in climates with very low humidity, this phenomenon would work much better than in damp climates.

These simple experiments seemed to imply that the static electric charge was something like the poles of a magnet in which unlike poles attracted each other and like poles repelled each other. It was speculated that there must be two different types of "forces" involved. Later scientists learned how to store the static electricity and build up a larger charge. Still later, scientists determined how to produce a flow of electricity rather than just storing up an at-rest electric charge. Whatever electricity was, it exhibited some type of "excess" as well as an equal "deficiency" in nature. A general misconception still existed that there was some type of spiritual force involved with static electricity and magnetism.

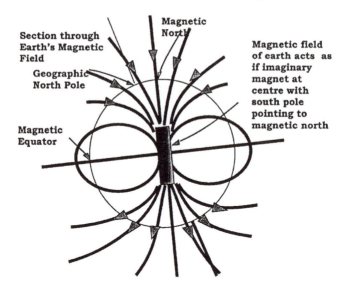

Figure 4.2 William Gilbert was an English physician who studied magnetism. He discovered that the Earth acted like a big bar magnet with magnetic poles and that a field of magnetic force surrounds the Earth.

Scientists of Electricity and Magnetism

William Gilbert (1540–1603) rediscovered what the ancient Greeks knew about rubbing substances to produce friction, which released electrical charges from solid matter. Gilbert is credited with distinguishing between the power of magnetism and the force of electricity. He coined the term *electric,* which means electron or amber in Greek. In addition to the term *electric,* he also used the terms *electric attraction, electric force,* and *magnetic poles.* Gilbert was the first to experimentally demonstrate that the Earth exhibits magnetic properties (see Figure 4.2). Although Gilbert's ideas predated much that would be explored later, he held the misconception that there must be some unseen spiritual force that causes magnets and electric charges to behave as they do.

Otto von Guericke (1602–1686) is best known for developing a vacuum inside two closed metal hemispheres that could not be separated, even with the strength of a team of horses, because of the normal air pressure on the outside of the hemispheres. He also developed the first electrostatic generator. It was made of a ball of raw sulfur that was rotated by a crank to produce a charge that could be discharged by touching it. He had no explanation for this phenomenon. His misconception was that all electricity was "static," meaning that it could build up a charge that could be released, but was not a continuing current.

Pieter van Musschenbroek (1692–1761) and Ewald Georg van Kleist (1700–1748) are both credited with inventing the Leyden jar, which was a primitive storage device for static electricity. The Leyden jar was at first called a condenser of electricity because it stored up or condensed an electrical charge. In modern electronics it is called a capacitor. This jar was named after the University of Leyden, located in Holland. The Leyden jar contained a metal cup suspended and insulated by silk cords placed in a glass jar. The cup contained water; a cork sealed the top opening. A brass wire passed through the cork, which touched the water. The jar was charged through the wire; the experimentors were not aware of the stored static electricity. When the jar was touched, it produced a strong shock. This is considered the first discharge of stored, artificial "lightning." Musschenbroek and van Kleist also mistakenly attributed this phenomenon to some unseen spiritual force.

The Leyden jar was greatly improved as a capacitor if metal foil was placed on both the inside and outside of the glass jar. This allowed the glass to act as a dielectric or nonconducting substance to separate the charges. A charge of stored static electricity occurred as the wire touched the inside foil, which was fed through the cork on the top of the jar. A circuit was completed when the wire conducted the electricity to the foil on the outside of the jar, or a spark jumped to your finger if it was brought near the wire exiting the jar. The same principle is used today in the Van de Graaff generator, which generates as many as 15 million electron volts inside modern particle accelerators.

Charles Bois-Reymond DuFay (1698–1739) was the first to make a distinction between the two types of electric charges obtained by rubbing different substances with different materials. He used rock crystals or glass rubbed with hair or wool to produce positive charges, and resin (amber) rubbed with silk or paper to produce negative charges. DuFay's main contribution was that he was the first to experimentally demonstrate that there are two, and only two types of electrical charges. He is credited with naming them *positive* and *negative*. He used the electroscope to measure the strength of an electric charge, which led to Coulomb's work with electrostatics. His misconception was that he called this a "two-fluid" system, which was later reinterpreted by Benjamin Franklin as a "one-fluid" system of electricity, which was also a misconception. Electricity does not flow like a liquid. One electron passes its charge to the next electron in a wire, but individual electrons do not "flow" through the wire as a fluid would.

Benjamin Franklin (1706–1790) is known for his experiments with static electricity as well as for his work in publishing and statesmanship. He also named the two types of charges for electricity as positive(+) and negative(–). He added the "+" sign to indicate excess charge, and the "–" sign to indicate a deficient charge. He also compared these two poles with the magnet's north and south poles. One of his major misconceptions was that electricity moved like a fluid. Franklin, and many others analogously compared electricity to water. Franklin's misconception was that he thought the greater quantity of a single "fluid" of electricity was stored in the positive pole, and it would flow to the negative pole because there was a lack of fluid at the negative pole. Franklin and others also held the belief that all materials held this "fluid" and that it could penetrate matter because it was neither lost nor created—just transferred. Later, when the electron was discovered, it was determined that electrons' changes "flow" from the negative pole to the positive. This "flow" may be thought of as the electrons pushing their charges along the surfaces of one atom to another in the conductor or wire. This concept of "flow" from negative to positive is just the opposite of what Franklin and most other scientists of his day believed, and as accepted by modern electricians.

These experiments led Franklin to speculate about the nature of lightning. In 1752 he flew a metal-tipped kite in an electrical storm. It attracted some of the lightning, but not directly (which was lucky for Franklin). A small charge of electricity followed the wet kite string to a metal key attached near his hand. When he brought his knuckles close to the key, sparks would jump from the key to his hand. This demonstrated that lightning is similar to artificially produced static electricity. It was also a very foolish and dangerous experiment. Several people who tried to duplicate Franklin's kite experiment were electrocuted. Benjamin Franklin also invented the lightning rod, which is used to protect structures from the devastating effects of the atmospheric discharge of electricity.

Joseph Priestley (1733–1804) (see also the section on pneumatic chemists in this chapter) knew from past reports that electrical charges become less strong with an increase in distance. He determined experimentally that the force of an electrical charge between two charged bodies (+ and −) diminished with the square of the distance between the two. He thought this was too simple an explanation, so he did not advocate it. He also determined that the charge on a body is evenly spread over its surface. His major misconception was that he believed electricity was still a "spirit-like fluid."

Charles A. de Coulomb (1736–1806) invented a special sensitive balance to measure the exact force exerted by opposite electrical charges. This led to the basic law of electrostatics, now known as *Coulomb's Law*. The Coulomb, or simply the symbol C, is the standard for measuring electrical charges. It is considered a large charge. One average-size lightning bolt has a charge of about 30 to 40 coulombs. Coulomb's major misconception was that he still advocated the two-fluid concept of electrical charges. Coulomb's major contribution was the *torsion balance* he invented, which was sensitive enough to measure very small forces and tiny weights.

This is about as far as the experimentation of static or at-rest electricity developed during this time period. Scientists eventually began to explore a different, more promising, form of electricity: the current. These scientists include an inquisitive physician and several chemists interested in the electrical nature of chemistry.

Luigi Galvani (1737–1798), a physician, is noted for his misinterpretations of experiments meant to demonstrate "animal electricity." First, Galvani stimulated the muscles of frogs with static electricity. The muscles jumped. He then wondered if lightning would cause muscle contraction. He connected wires to a frog's leg muscles and brass hooks connected to the spinal cord, while the frogs were resting on an iron railing. He placed the apparatus outside a window in a thunder storm and, when the static electricity of lightning was introduced, the muscles contracted. For some reason they continued to contract after the storm. This is an excellent example of how a unique experiment was set up incorrectly and produced a misconception related to the original hypothesis. Galvani called this delayed reaction *animal electricity,* which was disproved by Volta.

Alessandro Giuseppe Antonio Anastasio Volta (1745–1827) invented the voltaic pile, which is considered the first electric battery composed of several cells. His original group of cells consisted of a series of bowls containing salt water, each connected with alternating bridges of copper and zinc. It was a messy affair, but it did produce a continuously electric current. He improved his "battery" by alternating two different metal plates (copper or silver and zinc) separated by thick paper moistened with salt water as an electrolyte. When wires were placed on the top and bottom metal plates, a continuous current was produced when the circuit was closed. When cells are connected together in a series, they form a battery that produces a larger voltage than that of a single cell. He disproved Galvani's theory that animal muscle and nerve tissues contained an "electric fluid" by demonstrating that electricity did not come from the muscles, as Galvani asserted. Rather, it originated from the chemical reaction of the different metals used (brass and iron hooks) to connect the wires to the muscles. Of course, it was later proved that electricity is involved with live muscle and nerve tissues, but not as Galvani postulated. The unit of force of a "flowing" electric current is called the *volt* after Volta. The number of volts represents the degree of "pressure" or push behind the electric current (potential difference).

Michael Faraday (1791–1867) was both a chemist and physicist. He is credited with establishing a relationship between magnetism and electricity. He incorrectly believed that an ether or invisible substance in space was required for both magnetism and electricity to exert their forces. We now know that both magnetism and electricity are capable of exerting their forces in a vacuum or empty space.

Faraday is better known for his experiments in electrochemistry, which he learned about from his teacher, Sir Humphry Davy. His later experiments in electrochemistry led to what is now known as *Faraday's law of electrolysis.* Faraday is credited with coining the terms *electrolysis* (Greek for "loosen by electricity"), *electrolyte* (a solution that carries an electric current), *anode* (high road or positive charge), and *cathode* (low road or negative charge). He followed up some experiments performed by Hans Christian Oersted and André Marie Ampère, who discovered that an electric current produces a magnetic field. Faraday's idea was that if electricity can produce magnetism, why can't magnetism produce electricity? He used the idea that an electric current can be expressed as the number of lines that are cut in the fields of force by the wire intersecting these lines in the magnetic field. He demonstrated this concept in 1831, which has since became known as *induction.* It is the basis of the dynamo, or as we know it, the electric generator, which has revolutionized our way of life. Faraday's major misconception was his belief that electricity was a unifying force in nature and was responsible for chemical reactions, light, heat, magnetism, and everything else. He did not follow up his concept, which may have led him to explore the nature of electromagnetism. Faraday might be called the father of modern electrical devices.

Several other scientists of this period who made contributions to electromagnetism will be briefly considered.

Hans Christian Oersted (1777–1851) discovered that the magnetic needle of a compass was deflected at right angles to a wire carrying a current that was placed over the compass.

André Marie Ampère (1775–1836) advanced Oersted's discovery by demonstrating that the direction of the magnetic field is reversed if the direction of the current in the wire crossing the field is also reversed. He defined the unit for measuring the flow (current intensity or amount) of an electric current. It is called the *ampere* or *amp.*

Georg Simon Ohm (1787–1854) determined the amount of resistance there is to the "flow" of an electric current in different types of materials. The unit of electrical resistance is called the *ohm.* Ohm's law states the relationship between the flow or amount of current (amps), the pressure driving the current (volts), and the resistance to the flow of the current (ohms) in a closed circuit: (I [amperes] = V [volts] ÷ R [ohms]), or for volts it can be written as $V = I \times R$, or $R = V/I$ for resistance.

Joseph Henry (1797–1878) discovered electromagnetic induction before Michael Faraday. But Faraday published his work on electromagnetically induced currents first. Henry is given credit for the concept of self-induction, in which a magnetic field produces an electric field. He also designed the first electric motor to use brushes, developed powerful electromagnets, and invented the electromagnetic telegraph. His principle of induction is used in the small, brushless motors that drive the hard disks in computers and other equipment.

Albert Abraham Michelson (1852–1931) and Edward W. Morley (1838–1923) made a discovery equal in importance to Planck's Quantum Theory and Einstein's special and general Theories of Relativity. In fact, Michelson and Morley's discovery of electromagnetism may have led to the development of these and other theories. Their classic experiment was the downfall of the commonly held belief of the existence of an absolute limited space filled with an undetected matter called ether. The common misconception was that as the Earth moves through this ether, light, when measured on Earth, would have its speed altered according to the direction of the light in relation to the observer, who is also moving in the direction of the spin of the Earth. They developed a unique device called an interferometer that consisted of two mirrors at right angles to each other that reflected light back to a partially silvered mirror that splits the beam of light. If there was really matter (ether) in space, the beams of light would shift. Their experiment, repeated many times by others, indicated that it made no difference what position the beam of light or the observer were in relation to each other. Light always travelled at the same speed no matter where one was while observing the light beams or how fast the observer was moving. In other words, visible light reacted more like waves than particles. This concept was later expanded and proven by Einstein as his Theories of Relativity.

The discovery of the relationship between light and magnetism led to a new phase of physics, sometimes called *unification physics* or *unified field theory*. Even today, scientists attempt to unify, in a single equation, several forces; for example, the TOE or GUT Theory is expected to include *all* radiation and *all* nuclear particles, gravity, plus the strong and weak nuclear forces, and so on. The unification of electricity and magnetism led to many new discoveries (and misconceptions) in other fields, including optics, chemistry and, later, particle physics. Let us now consider some of the rapid discoveries and developments in chemistry—the sister science of physics.

The Golden Age of Chemistry

Chemistry involves the outer portion of atoms and molecules where the chemical (energy) action between elements and compounds takes place. Although the nucleus of the atom currently is considered part of the realm of physics, this was not always the case. The invention of instruments designed to accurately measure what goes on in chemical reactions parallels the developments of chemical theories, or it may be thought in the reverse; that is, that the need for better instruments followed the development of theories in chemistry.

Scientists of the Golden Age of Chemistry

Robert Boyle (1627–1691) was previously mentioned in this chapter in the section "Bacon's Contemporaries." He is important because he was one of the first to come up with new ideas about matter, particularly air and gases. Boyle may be thought of as the first of the pneumatic chemists. He proposed that all substances contain elemental or unmixed bodies and that these bodies can be compounded. Because Boyle was a "corpuscularian," his misconception was that he believed that the characteristics of different bodies could be explained by how much or how many basic corpuscles they contained. His corpuscles might be equated with atoms, which were not then known. He also believed that corpuscles can combine to make compounds and compounds can be separated into their corpuscles.

Boyle observed that heated metal gained weight. His misconception was that the weight gain was the result of the absorption of a substance he called "igneous particles," which could pass through just about anything. He explained combustion and respiration as caused by his "three types of air particles," one of which was in evidence in air. The other two air particles came from the Earth and space, but only in minute amounts. His demonstrations proved that certain things will burn in air, but others, such as gunpowder, will burn under water, so he correctly concluded that other substances besides air can support combustion. But he held the false belief that some form of sulfur must be present in all materials that burned. Boyle's famous gas laws convinced others of the uniformity of air. He stated that the pressure and volume of a gas will increase if the temperature of the gas is raised, or the pressure and volume will remain unchanged if the temperature remains constant. This concept stimulated pneumatic chemists to examine chemistry in a more quantitative than qualitative manner. Boyle's work led Priestley and Lavoisier to further developments in chemistry.

Johann Joachim Becher (1635–1682) and his student, Georg Ernst Stahl (1660–1734) used Boyle's ideas to develop the *phlogiston theory* (see also the section, Bacon's Contemporaries, in this chapter). They held several major misconceptions. First, they believed that all bodies were composed of three basic substances, mainly three types of earth: vitreous, mercurial, and fatty. Second, that all combustible substances were mainly "fatty" substances that somehow disappeared when burned. Stahl consolidated these concepts into the phlogiston theory, which in the days of qualitative chemistry was well accepted because it did answer many questions about combustion.

Stephen Hales (1677–1761) invented a new way of collecting gases by inserting a tube into a container whose mouth was under water (the pneumatic trough). He conducted many combustion experiments, but arrived at many incorrect conclusions. He neglected to account for the solubility of a gas as it passed through the water to the collection chamber. Thus he arrived at incorrect volumes for several gases. He discovered several poisonous gases. His major misconception was his belief that the characteristics of an "air" or gas were fixed in the substance from which it was derived, not in the particular gas itself. He called this characteristic "true air." His major contribution was his work in the respiration and transpiration of living plants.

The Pneumatic Chemists

Joseph Black (1728–1799) was one of the major pneumatic chemists. Contrary to Hale's assertions, Black determined that the characteristics of a gas were unique to itself and not to the source of its substance. In other words, he changed the then prevailing idea that all gases were parts of air. Black conducted many experiments with magnesia and quicklime and produced what he called "fixed air," which demonstrated that gases can take part in chemical reactions. Although he is credited with using quantitative terms for his experiments, some of his conclusions for his fixed air were not clear, particularly when other gases that had not yet been discovered contaminated his results. He did demonstrate that a small amount of fixed air existed in the atmosphere, that fermentation produced fixed air, and that when humans exhaled, they produced fixed air, which later turned out to be carbon dioxide.

Henry Cavendish (1731–1810) identified "fire air" in 1766 as a product of the reaction of acid on metal. In Boyle's time this gas was known as "inflammable air" but had not been identified. We now call this gas *hydrogen*. Cavendish weighed specific volumes of different gases to determine their densities. He found that his fire air was the lightest of all gases then known, but also highly combustible. Cavendish demonstrated that when two measures of his inflammable air and five measures of common air were exploded, no inflammable air remained in the vessel. However, the total volume of common air was reduced by one-fifth, and a small amount of colorless liquid was produced.

Joseph Priestley (1733–1804) was an amateur pneumatic chemist who studied more gases than anyone up to his time. He was also a phlogistonist who mistakenly believed that all gases contained phlogiston, even when he discovered some gases that did not fit this description. Phlogiston comes from the Greek for "to set on fire" and was coined by George Ernst Stahl to incorrectly describe combustion. Priestley did not like to weigh and measure things but rather preferred to describe their characteristics in detail. He developed new types of equipment and techniques that were used by other chemists. He heated mercuric oxide to produce a gas that aided combustion and respiration. This gas was later identified as oxygen by Lavoisier. Because this new gas did not have the same properties as his phlogistic gases, Priestley named it "dephlogisticated air." His misconception was that his new air had somehow lost its phlogiston and was now ready to absorb phlogistic air from animals and the combustion of fuels. Priestley added fixed air (CO_2) to water to make soda water. If you add flavoring to his carbonated water, you could say he invented soda pop.

Carl Wilhelm Scheele (1742–1786) was one of the last of the pneumatic chemists. He was also a phlogistonist. He made many contributions to the advancement of chemistry, but his big misconception, like so many of his contemporaries, concerned combustion. He believed that because combustion can take place in air, that the air itself must be analyzed to explain combustion. He believed that the nature of fire differed for each particular substance that burned. Scheele believed that when air combined with the phlogiston of a substance, the phlogiston escaped from that substance. He held this belief because the substance itself was reduced as it burned.

Founders of Modern Chemistry

Antoine-Laurent Lavoisier (1743–1794) is considered one of the fathers of modern chemistry. Lavoisier did not discover any new elements, but rather designed and synthesized a new system of chemistry that is still used today. He published a comprehensive textbook that described his system of elements, which was based on the conservation of mass in chemical reactions. Up to this time, many scientists believed that mass was lost during a chemical reaction. Although Priestley discovered oxygen, he did not know what it was. He named it "dephlogisticated air." Scheele called it "empyreal air." Black confused it with his "fixed air." Cavendish included it as a type of his "inflammable air." Lavoisier, at first, named it "highly respirable air," and later he called it "vital air" because it supported life. Lavoisier later gave it the name *oxygen,* which is Greek for "acid-former," because of the misconception that all acids contained oxygen.

Lavoisier's many experiments with oxygen led to the demise of the phlogiston theory held by Stahl and others. He demonstrated how oxygen was chemically involved with combustion, respiration, and rust (called "calcination" in those days). Lavoisier held several major misconceptions: First, he believed, as did his contemporaries, that oxygen was essential to all acids and was part of an acid's composition. Second, he believed that heat was a chemical element—not a physical property. Third, he was confused about combustion. He believed that when something burned or rusted it gained weight because it combined with Black's fixed air.

A major experiment conducted by Lavoisier was the decomposition of water, from which he theorized that the recomposition of oxygen and hydrogen would form water. He said that this proved that water was not an element but a substance. This concept was later proved correct by Gaspard Monge (1746–1818) when he burned a measured amount of hydrogen in a measured amount of oxygen, which formed water that weighed about the same as the combined weights of the two original gases. Many scientists of this period accepted Lavoisier's concepts of quantitative analysis, whereas others, including Priestley, supported the phlogiston theory until their deaths.

This new chemistry led to a rethinking of the atomic theory of matter first proposed by the ancient Greeks. Many scientists from various countries contributed to atomic theory. Some of the most noteworthy of these scientists are Klaproth of Germany, Vauqelin of France, Wollaston of England, and Higgins of Ireland, as well as Bergman, Berthollet, Wenzel, Richter, and Proust.

William Higgins (1766–1825) devised a chemical system based on forces that held particles together (later known as valence and bonding energy) and a rule of multiple proportions that preceded the work of John Dalton. Higgins's misconception was that he believed atoms could combine only in certain proportions by volume, not weight. Credit for his work was given to John Dalton.

John Dalton (1766–1844), was the second giant of modern chemistry. He built his concepts of chemical atomism on the work of the ancient Greeks Leucippus and Democritus. Dalton built on Higgins's and Wollaston's concepts of *multiple proportions,* which he later refined as his *law of multiple proportions.* He worked with the physical behavior of gases, which soon led him to his major contribution: atomic weights.

This was a rather elegant theory for its time. It was based on the relative weights (not volumes) of the known elements as compared to the lightest element, hydrogen, which was used as a base and assigned the relative weight of 1.0. His theory was based on accepted assumptions of elements composed of indivisible particles called *atoms.* All the atoms of one element are the same, and chemical reactions occur when different types of atoms of different elements combine to form a new substance. A misconception that was also quite confusing was that this new particle of combined atoms was also called an *atom.* This confusion continued until the term *molecule* was introduced to represent these new types of particles, which consisted of joined atoms of the same or different elements; for example, O_2 and CO_2. Dalton developed a Table of Weights of Elements that included over 20 elements, gases, acids, and compounds. Because the concepts of valence and chemical formulas were not yet developed, he made many mistakes in applying his theory. For instance, he believed that particles of all gases were like buckshots that varied in diameter. Thus, equal volumes of different gases did not contain equal numbers of particles, which is only true if the proportion of each gas's atomic composition is known. Because the confusion of combining volumes with weights of elements (particularly gases) persisted, it was sometime before these relationships were understood.

The Gas Laws

Sir Humphry Davy (1778–1829), Jöns Jakob Berzelius (1779–1848), Michael Faraday (1791–1867), Amedeo Avogadro (1776–1856), Jacques Alexandre César Charles (1746–1823), Joseph Louis Gay-Lussac (1778–1850), and many others contributed to what are now known as the *gas laws*. The gas laws use both physics and chemistry to explain the behavior of gases under a variety of conditions. The temperature, pressure, and volume of a gas determine how that gas will behave. As mentioned, Boyle's law states that the volume and pressure of a gas will be maintained if the temperature does not change. Charles's law states that the volume of a gas is inversely proportional to the absolute temperature of the gas, if the pressure remains the same. Gay-Lussac's law is similar to Charles's law, which states the relationship between the temperature and volume of gases. Gay-Lussac's law states that the pressure of a gas is inversely proportional to the gas' absolute temperature, if the volume of gas remains the same. These laws are combined to form the generalized "ideal" gas law. It can be stated as $pv = nRT$, where p = pressure of the gas, v = volume of the gas, n = number of moles of the gas, R = the gas constant, and T = absolute temperature of the gas measured in the Kelvin scale. This law applies to most gases at normal temperatures, but not at extreme temperatures.

The utility of the ideal gas law is that it enables chemists to determine the weight of a gas (which is difficult to measure directly) if they know its volume at a given pressure and temperature. There was a general misconception about this law, which assumed that it would hold up for all temperatures from absolute zero (–273°C) to very high temperatures. When the law was formulated, the exact motion of molecules at very low and high temperatures was not known. Because gas molecules slow down at very low temperatures and become plasmas at high temperatures they either stay close together or far apart and do not have the same volume relationships that they have at room temperature. In theory, at absolute zero all molecular motion stops or the molecules just vibrate in place.

Scientists Leading into the Twentieth Century

The latter part of the eighteenth century through the nineteenth century saw many corrections in the field of chemistry. Standard nomenclature, formulas, equations, and measurements were developed, accepted, and used. New and improved instruments were developed. One of the most important advancements, which allowed the identification of new elements and the determination of their characteristics, was a series of spectra instruments that could analyze electromagnetic radiation emitted or absorbed by elements.

Each element, when heated or excited, has a very unique, measurable electromagnetic spectrum. Once the identifying spectrum of a specific element was determined, it was comparatively simple to use a spectroscope to compare the spectra of unknown substances and accurately identify their elements.

During this period, a few scientists helped pave the way into the twentieth century. Their work enabled the discoveries of those who followed.

Dmitry Ivanovich Mendeleyev (1834–1907) recognized the work of several others who saw some relationship between the atomic weights of different elements. He recognized the periodic or recurring nature of characteristics of elements based on their atomic weights. The elements' octet nature (characteristics repeat after every eight elements) of increasing atomic weights led to his periodic table of the chemical elements. Because neither Mendeleyev, nor other scientists of that time, were aware of the basic structure of the nucleus, the atomic weights were the best information available to formulate this table. It was obvious that several blank spaces appeared in the table. He attributed these blank spaces to unknown, yet-to-be-discovered elements. Over the next decades new elements were discovered that fit exactly into the blank spaces predicted by Mendeleyev in his periodic table. His misconception was that the periodic nature of elements was based on the atomic weights (total number of neutrons and protons in the nuclei, whose existence was not known), rather than on the number of protons (positive charged particles) in the nuclei of atoms or the equal number of electrons in the shells or orbits of neutral atoms. Using atomic weights instead of atomic numbers created several serious errors in the periodic table. It was later determined that the number of protons in an atom's nucleus determines that element's atomic number—not the atomic weight of the atom. Once the periodic table was revised to represent the proton numbers, rather than the atomic weights of elements, it proved to be an exceptional conceptual schema that advanced chemistry (see Figure 4.3).

Joseph John Thomson (1856–1940) is credited with discovering the electron, the first basic subatomic particle to be identified. He pictured the atom as a positively charged "fuzzy ball" that had negatively charged electrons imbedded in it. This concept, although incorrect, did explain the neutral charge of the atom. According to his theory, this neutral atom, under stress, could expel electrons that could "flow"—thus electricity.

Ernest Rutherford (1871–1937) is credited with coining the terms *alpha rays* (α), which have relatively heavy positive radiation (later known as helium nuclei), and *beta rays* (β), which are much lighter in weight and carry a negative charge (later known as high-speed electrons). Both alpha and beta rays, as well as *gamma rays* (γ) come from the nuclei of uranium and other sources of radioactivity.

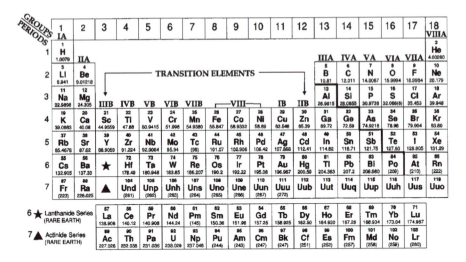

Figure 4.3 The modern periodic table of the chemical elements. This table is arranged in octaves (repeating characteristics for every eight elements) by the atomic number (protons in the nucleus) rather than by atomic weights as in the table originally devised by Dmitri Ivanovich Mendeleyev.

Rutherford's classic experiment consisted of "shooting" alpha particles at a very, very thin sheet of gold foil (about one fifty-thousandth of an inch thick). Most of the alpha particles went through the foil as if nothing were there and were detected by a photographic plate placed behind the foil. But some alpha particles were deflected to the sides and others bounced back to the source. His conclusion was that atoms are mostly empty space with a very small, positively charged center that had a great deal of mass. If all the nuclei of all the atoms that make up your body were consolidated into a single ball, the ball would be smaller than the size of a period in this text. It was evident that this massive, small positive center did not take up much of the total space of the atom. This experiment led to his theory of the nuclear atom with its orbiting electrons, somewhat like a miniature solar system. We now know that the nucleus takes up only about one one-hundred thousandth of the total diameter of an atom. This is another example of the inseparable nature of physics and chemistry.

PARTICLES, COMPLEXITY, CHAOS, AND UNCERTAINTY

During the first part of the twentieth century, the theories of matter, energy, space, time, and the relationships between chemistry and physics were consolidated. The second half of the twentieth century (after World War II) was a time when ideas were refined, and the accuracy of measurements fine-tuned, and more accurate theorizing about the unknown was possible.

Analytical Chemistry

In the late nineteenth and early twentieth centuries analytical chemistry demonstrated that atoms, although existing in all elements, are unique to each specific element. Analytical chemistry also showed that atoms of elements combine in a very exact way in exact amounts to form new substances (molecules of compounds) that bear no resemblance to the original combining elements. Everything in the universe is composed of only about 100 specific and identifiable types of atoms arranged in a great variety of forms (radicals and molecules). Dalton, whom we described as one of the founders of modern chemistry, pointed the way for our understanding of chemistry. Even so, one of his major misconceptions was his failure to account for the law of multiple proportion as it distinguishes between atoms and molecules. The law of multiple proportion states that an element of one mass may combine with another element with a different mass in more than one way, based on their atomic weights; for example, 12 grams of carbon will combine with 16 grams of oxygen to form CO (3:4), whereas 12 grams of carbon will also combine with 32 grams of oxygen to form CO_2 (3:8). The multiple proportions of oxygen in the two different molecules is 16:32, or 1:2. It had been demonstrated that hydrogen and oxygen, when combined by combustion, formed water. But Dalton and some other chemists could not determine if the ratio of hydrogen to oxygen was 1:1, 2:2, or 2:1.

Valence

Three nineteenth-century scientists came up with concepts that led to the idea of valence, which is based on interpretations of the electron's electrical charge and mass, as determined by J. J. Thomson.

Edward Frankland (1825–1899) observed how metals combined with organic compounds. He considered this bonding to occur in whole numbers, which he called *valence,* meaning "power" in Latin. His misconception was that he believed that valence was related to an atom's atomic weight, rather than the number of electrons in the outer orbit (or atomic number based on the number of protons in the nucleus).

Alfred Werner (1866–1919) developed a coordination theory that led to his concept of valence. He studied the affinity between atoms by using the attractions of their electrons and protons. His misconception was that he thought all atoms combine into molecules in a one-to-one manner.

Nevil Vincent Sidgwick (1873–1952) picked up on Werner's work and developed the *electronic theory of valence,* which contributed to our understanding of chemical bonds in chemical reactions. At this time he neglected to recognize what we call *radicals,* which are groups of atoms in a compound that also has an electrical valence, and must be considered in order for the valence of neutral molecules to equal zero.

The Structure of the Atom

Niels Henrik David Bohr (1885–1962) is considered more of a physicist than a chemist, possibly because he was the first to apply the rather new *quantum theory* to the structure of the atom. Bohr's concept of the atom corrected some weaknesses in the Rutherford atom. Rutherford's atomic model mistakenly stipulated that an atom's electrons would continually radiate energy as they moved around the nucleus. If this occurred, all the electrons would soon loose their energy and be attracted to the positive central nucleus, which might result in the end of matter! Rutherford's model of the atom was more like the solar-system model, whereas Bohr used quantum theory as an energy concept for his model of the atom, which was known as *quantum mechanics.*

The *exclusion principle,* developed by Wolfgang Pauli in 1925, stated that two electrons cannot occupy the same quantum or energy state in the same orbit of an atom at the same moment. As long as the input of energy continued, electrons "jump" from inner orbits to outer orbits until they reach and escape the outermost orbit, thus changing the neutral atom into an ionized particle (an atom with an electrical charge) with a positive electric charge, and a free negative electron. This jump is what we call a *quantum leap* of energy. Many people have the misconception that a quantum leap is a great jump or distance when, in fact, it is just a mathematical concept to explain a very, very small change. Quantum theory is one of the most basic laws of science.

Bohr and others refined the solar-system analogy for the structure of the atom to include subshells, this became necessary when it was realized that not all the electrons in some orbits had the same energy levels. This solar-system analogy for atomic structure was a misconception because it would have to include gravity, and gravity as a force is much too weak to be of any consequence between the particles that make up atoms. Thus the electrons are not maintained in their orbits by gravity as are the planets around our sun. Rather, the theory of quantum mechanics governs electrons in their electric fields (energy and momentum).

Bohr and others recognized several major misconceptions or limitations in his model. Most of these were in his model's applications to the heavier and more complex atoms that had numerous electrons in several orbits. The mathematics and quantum mechanics worked just great for the single electron in the hydrogen atom, which contains a single proton in the nucleus.

This limited hydrogen-atom model became known as the *standard model of elementary particles and forces.* But the heavier atoms were much too complicated to calculate with this system. Atomic structure became more understandable when related to energy; thus atomic structure is more in the realm of modern particle physics than chemistry, and is known as *physical chemistry.*

Quantum Theory

Quantum theory grew out of the discovery of what is called black body radiation, which was discovered by Gustav Robert Kirchhoff (1824–1887) and was further developed by Max Planck.

Max Karl Ernst Ludwig Planck (1858–1947) produced an equation based on a theory that explained the nature of the radiation emitted out of a small hole from a heated black metal ball, usually referred to as *black body radiation.* Planck's equation postulated that light was not a continuous stream of particles. Rather, a beam of light is a series of many, many very small discreet pieces (packets or chunks) of energy. Each unit is called a *quantum.* Furthermore, he calculated that the shorter the wavelength, the larger the chunk and thus the greater the energy of that bit of light. He called these tiny pieces of energy *quanta,* which is Latin for "how much." The mathematical relationship between the wavelength and energy of a particle became known as *Planck's constant,* expressed as *h,* or when expressed as the energy of a quantum of light, $E = h\nu$, which is an extremely small amount of energy. Planck's constant has proven to be one of the most useful of over three dozen scientific and mathematical constants, and it is used to explain many phenomena in science. Planck had running battles with many scientists of his day, but his theory led to the end of Newtonian classical (mechanistic) physics as it ushered in Einsteinian (relativity) physics. Planck is credited with saying that new scientific truths do not gain acceptance through the convincing of one's opponents, but rather new ideas become acceptable when one's opponents die and a new group, familiar with one's theories, become supporters. Planck's misconception was that he believed that light was of a corpuscular nature rather than pure electromagnetic energy with momentum, as proposed by others. Later, his quantum theory was applied to both the wave function of electromagnetic light and the particle momentum of the electromagnetic photon, which acts as mass in Einstein's new concept of gravity.

Relativity

Albert Einstein (1879–1955) is generally credited with the final dethroning of Newtonian physics as he introduced the twentieth century to the new physics of *relativity*. He developed two theories of relativity. The first he called the *theory of special relativity*, which dealt with aspects of space/time related to motion. The second was his *theory of general relativity*, which included gravity. Both are somewhat difficult for us to understand because in everyday life we deal with very limited distances and velocities (speeds) on Earth, which can be adequately explained and understood by the use of Newtonian mechanistic physics. The reason Einstein's physics is called *relative* is, according to the theory, that for very great distances and velocities in the universe there is no such thing as *absolute* rest or motion. Also, absolute space and time are dependent on velocity. Einstein used the quantum theory to explain that light consisted of tiny energy packets (quanta), which he later called photons. These light quanta had energies that exhibited frequencies that were inversely proportional to their wavelengths. According to the theory of relativity, nothing can travel faster than the speed of light. An infinite amount of energy (which is finite in the universe) is required to accelerate a body or a particle with mass to the speed of light, which is obviously an impossibility because there is only a limited amount of energy in the universe.

In 1915 Einstein expanded his ideas into a general theory of relativity, which in part is based on the curvature of time and space and took into account his new concept of gravity based on momentum. He altered the concept of three-dimensional (length, width, and depth, or X, Y, and Z) space to four dimensions by adding time to the other three dimensions. This was truly revolutionary. The fourth dimension is described as a combined nonabsolute, curved space–time, which is based on the absolute constant speed of light. The general theory states that the speed of light is the same for all observers, *regardless of their frames of reference*. The speed of light is expressed as a universal constant *c*, which is about 186,000 miles/second. A famous 1919 experiment proved Einstein's prediction that light from a distant star would be bent by the gravity of the sun as it passed by the sun's edge. His theory stated that light, as an electromagnetic field with energy and momentum but no mass, would be affected by the sun's gravity. This is exactly what happened to the light emitted from *several stars* viewed during an eclipse of the sun, thus establishing the viability and acceptance of his general theory of relativity.

Einstein's theories proved to have some weaknesses, and he held several misconceptions that later in life isolated him from the scientific community. His general theory did introduce the field of cosmology, which is the study of the universe. He also was correct in explaining that gravity would prevent light from escaping from very, very, dense stars that collapsed into a pinpoint. In 1997, 80 years after Einstein made this theory, new observations confirmed that spinning black matter and neutron stars swirl like water in a drain, thus making everything, including light, enter the whirlpool, never to escape. Thus they would become "black holes" and possibly new universes. He also theorized that large bodies moving in space generate a force similar to the electromagnetic force around a wire carrying electrons. Thus this force could move around. Even so, his major misconception was that the universe was static in the sense that it was neither expanding toward infinity nor shrinking toward a collapse into nothing. More recently, cosmologists proposed a nonfinite expanding model. Others proposed a regeneration model in which new matter is constantly produced as the universe continues to expand into infinity. Still others propose an oscillating model that will expand, then contract to nothing, and possibly start all over again as a new big bang. Einstein was also deterministic. He was not religious in a classical sense, but he believed that "Quantum mechanics is very impressive. . . . God does not play dice" (Asimov and Shulman, 1988, p. 222). This might be interpreted to mean that nature is not random, but rather follows a plan imposed by a superior entity. As we will see later, chance and randomness are useful concepts, particularly when considering some recent theories. Einstein did not accept the ideas and developments of several other physicists. He never completely accepted Heisenberg's own quantum theory, or Heisenberg's concept of indeterminacy. Heisenberg removed Einstein's absolute determinacy, or cause and effect, from physics and replaced it with a mathematical/statistical probability. This concept, though extremely important, did not neatly fit with Einstein's theories of relativity.

Quantum Mechanics

Werner Heisenberg (1901–1976), Max Born (1882–1970), and Erwin Schrödinger (1887–1961) are credited with advancing *quantum theory* into the science of *quantum mechanics*. By using quantum mechanics, scientists were able to predict many things in physics, including the position and energy of orbiting electrons. There are limitations to using the short wavelengths of electromagnetic sources (light) to measure an extremely minute particle's exact position and momentum (mass × velocity) at the same time with any accuracy. This is the essence of Heisenberg's famous *uncertainty principle* (also called *indeterminacy*). The more precisely or accurately you observe a particle's momentum, the less exact will be your knowledge of its position, and vice versa. But statistical probabilities of a particle's amplitude (mass/position) and motion can be determined. Our current concept of quantum mechanics is that, in various forms, it controls everything in the universe. Heisenberg knew that the nuclei of atoms had to contain Sir James Chadwick's (1891–1974) neutrons as well as protons to account for an atom's atomic weight (mass). One of Heisenberg's misconceptions concerned the amount of energy required to maintain the close association of positive (thus repelling) protons in the nucleus. He neglected to consider the strong nuclear force that was required to prevent multiple protons in atomic nuclei from repelling each other. The other misconception was that he did not realize that for a neutron to become a proton, the neutron had to give up an electron as well as some radiation.

Matrix Mechanics

Paul Dirac (1902–1984) expanded and unified Planck's quantum theory, Schrödingers's wave mechanics, and Heisenberg's ideas of quantum mechanics into an improved theory of matrix mechanics. He used more precise mathematical descriptions of elementary particles by developing and arranging an array of numbers for the particles into a three-dimensional matrix, which was a new version of quantum mechanics now called *matrix mechanics.* One of his contributions was the idea that for every type of subatomic particle there must be an opposite counter particle with equal mass but opposite charge. He called these *antiparticles.* Dirac's main misconception was that he thought quantum mechanics could be combined into a theory of relativistic wave mechanics. A more correct version, now known a *quantum field theory,* combines the wave properties of particles with relativity.

Complexity

This brings us to the latter part of the twentieth century with its theoretical refinements, the discovery of numerous subatomic (elementary) particles, and the studies of complexity and chaos. Physics is basically the study of mass and force within the very small and very large dimensions of our universe. This, in essence, is a reductionist concept because it subsumes all other sciences and human activities under the rubric of physics. Reductionism assumes that only a few physical laws will ultimately explain everything in the universe. As Carl Sagan (1995, pp. 270–71) says, "Reductionism seems to pay insufficient respect to the complexity of the Universe. It appears to some as a curious hybrid of arrogance and intellectual laziness." Searching for the ultimate subatomic elemental (or rather subnuclear) particles is also seen as the search for the "theory of everything." Einstein attempted such a theory of everything, and tried to accomplish it with his grand unification theory (GUT) for gravity, mass, motion, and space/time. This search continues.

The particles that make up the atom and its nucleus are extremely tiny and, for the most part, are held together by extremely strong forces. Identifying these particles required the development of unusual instruments that were capable of separating them, identifying them, and analyzing them. Instruments that knock the particles out of the atom are called particle accel-

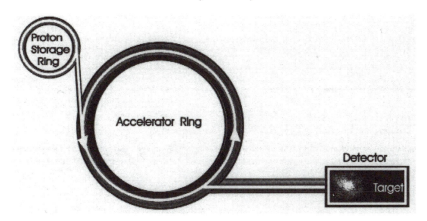

Figure 4.4 Diagram of a typical particle accelerator. Protons (or other subatomic particles) are sent into the ring where electromagnets force them to reach great speeds until they are directed to a target, which is usually another particle. The collision is detected and recorded at this stage.

erators, colliders, or just atom smashers, which can be either circular or linear in shape (see Figure 4.4).

Subatomic Particles

Following World War II many scientists began searching for new subatomic particles that might be useful in the continued understanding of the physics of nature. There are hundreds of physicists involved in this search for an ever-expanding list of subatomic particles. We will list just a few of the major particles and some of their related forces that have been found. It is difficult to identify specific misconceptions or mistaken beliefs that occurred during this period. It may suffice to say that many of the particles, when first theoretically conceived, produced, or observed were either incorrectly identified or rather misidentified until additional experimentation and confirmation was forthcoming. Steven Weinberg (1993) states:

In this century no theory that has been generally accepted as valid by the world of physics has turned out simply to be a *mistake,* the way that Ptolemy's epicycle theory of planetary motion or the theory that heat is a fluid called caloric were mistakes. . . . The consensus in favor of physical theories has often been reached on the basis of aesthetic judgments before the experimental evidence for these theories became really compelling. (p. 130)

There are two main classifications of subatomic particles: The *fermions,* which are "matter" particles (e.g., electrons, neutrons, neutrinos, muons, quarks, or leptons), and the *bosons,* which are "force" particles (e.g., bosons produced by quarks, gluons, photons, positive W and negative W particles, neutral Z particles, and the graviton).

The search for the ultimate particle and the ultimate unifying law of science continues. Physicists and other scientists are speculating, theorizing, experimenting, and philosophizing about these matters. Some are attempting to improve and become more precise about what we do know and extrapolate new directions from this knowledge. Others are branching out to travel new paths, looking for new directions, and answers to age-old questions. This brings us to two rather new fields that are not completely accepted as the path to take in this quest. The first is complexity.

The science of *complexity,* though not well defined, tries to answer questions that other sciences do not seem to be able to answer adequately. These questions, on the surface, may not seem to belong in physics, but then, ultimately, everything may be considered physics. A few examples: How did life begin? How did single living cells develop into higher organisms? What drives the evolution of living organisms, and of the universe? Why did the stock market crash in 1929? Why can't the stock market be accurately predicted? Is there a mind separate from the brain? What are sight, hearing, thinking, memory? Why are there so few different types of atoms, but so many particles in atoms? What causes chemicals to react as they do? Is valence a predetermined condition? What happens at the surface interface when water freezes, boils, evaporates? Why is weather so unpredictable? Why does entropy (disorganization) always run in one direction (toward randomness)? How can simple molecules form complex molecules of life if entropy tells us that the norm in nature is toward randomness and simplicity? What existed before the "egg" exploded as the big bang? Why did the big bang occur, and why does matter/energy continue toward random disorder, but yet form organized systems of all kinds (living organisms, galaxies, etc.)?

These and similar questions all have several things in common. First, so far they are all unanswerable. Second, they all deal with complex systems. And, third, similar questions just might be more adequately addressed by philosophy than by physics. Okay, so what are the characteristics of complex systems? First, they are made up of many (often very many) interacting parts that, in themselves, may also be complicated subsystems within the complex system; for example, the interaction of molecules of air, water vapor, and temperature in a storm system. Second, they are not just complicated, rather they are forms of disorder that exhibit a type of spontaneity. Third, complex systems have a built-in form of self-organization and cohesiveness, unlike chaos. Fourth, complex systems are never stable, they are always in transition—never in equilibrium (what form will a cloud take in the next 2 or 15 minutes? Can you predict the shape of a cloud?). And fifth, complex systems, being adaptive, can only evolve and "improve" as a system. They cannot be stable because they are always changing.

The origin of life and its evolution are good examples of a complex system. One concept is that a primordial soup contained the correct types of atoms and molecules, plus favorable physical conditions of atmosphere, radiation, energy, and so on that were required to start life. These compounds and energy are thought to have formed a self-reinforcing and coherent set of physical and chemical reactions. These conditions would allow for a spontaneous, self-organizing, cooperative system that was adaptive to its environment, thus producing more complex structures. These systems, once started, could grow and change as their internal structures mutated and became better adapted to external changes.

Chaos Theory

The jump from these primitive organic molecules to complex life forms is what makes complex systems interesting. A group of scientists have formed the Santa Fe Institute, a think tank for the study of complex systems. They have come up with several new concepts, including the spontaneous organization of complex systems and the renormalization theory, which describes the critical-phase transition between systems. Examples are what occurs between the two forms when water becomes ice, or water evaporates in the atmosphere and becomes a gas. Complexity draws on other areas of science; for example, information theory, computer science, thermodynamics, statistics, and mathematics, as well as biology and economics. There seems to be a sequence in all dynamic complex systems:

$$\text{Order} \Rightarrow \text{Complexity} \Rightarrow \text{Chaos}$$

A new science of *chaoplexity* combines chaos and complex systems.

One of their most interesting concepts related to complexity is *chaos theory.* According to ancient Greek mythology, chaos was the prebeginning of the Earth, the heavens and even the gods, when everything was "unbeginning." The chaos theory deals with the breakdown of ordered systems into random, disordered, or rather chaotic systems. Chaos can be found in many areas of science; for example, in the second law of thermodynamics, molecular motion, the theory of organic evolution, theories of human behavior, and economic theories. Chaos is a main theory in the new science of chaoplexity. A simple definition is that a chaotic system is very sensitive to initial conditions of that system. Small initial starting points, no matter how minute, combined with small changes along the way, will cause the system to progress in very different ways. Depending on environmental conditions, the system progresses differently in the early phases than it does in the final phases. James Trefil (1992) provides an excellent example of a chaotic system:

A chaotic system is one in which the final outcome depends very sensitively on the initial conditions. White water in a stream is a good example of a chaotic system. If you start a chip of wood at one position, it will come out at a particular point on the other side of a rapids. If you start the second chip of wood at a position almost (but not quite) identical to that of the first, the second chip will—in general—come out of the rapids far from where the first one did. The final outcome (the position of the chips) thus depends sensitively on the initial conditions (the place where they started their journey). (p. 183)

It is impossible to start the chips at *exactly* the same place each time. The behavior of the chips is unpredictable, possibly the effects of the environment on them is random, and the chips may end up where they do by chance. This is what chaos theory is trying to determine. Is there a pattern or "solution" to complex systems? Are there ranges of prediction and accuracy in the system? Are the solutions to these problems basically unknown mathematical solutions? Also, is there a transition phase between order and chaos as, for example, the interfaces between the states of matter—solid to liquid or gas, or liquid to solid or gas? This is referred to as the *edge of chaos.* Chaos also exists in turbulent systems such as whirlpools, the flow of fluids in pipelines, and long-term climate as well as short-term weather patterns, including cloud and storm formation. Chaos affects the flow of oil in a pipeline. You may have noticed chaos in the bunching up and spreading out of cars in the flow of automobile traffic on open superhighways. Chaos appears in dynamic systems such as long-range weather predictions, ecological systems, evolution, and population systems. One problem is that many phenomena in nature are nonlinear. Instead, they branch, seemingly chaotically. They are inherently unpredictable because of the many small influences that have an effect on the nature and progress of the complex system.

It is too early to tell if there are many misconceptions and false beliefs related to complexity and chaos theories. The studies are just too new, and certainly, these theories are not yet accepted by all physicists.

What does the future hold for chemistry and physics in future centuries? Scientists seem to have several different points of view when this question is posed. Some say that as we explore nature and continue to look for and find smaller subatomic particles and energies, and possibly the "final answer," that we will find that the universe is even less comprehensible and possibly even less meaningful to us as human beings. In other words, the final answer may have less and less to do with us, and more and more to do with yet-to-be-discovered answers to a very complex system that is the universe. And yet,

Einstein said, "The most incomprehensible thing about the word is that it is comprehensible" (Asimov and Shulman, 1988, p. 211).

Another point of view held by some is that science is basically finished, not necessarily the biological sciences, but at least the physical sciences. The belief that science knows all there is to know and has discovered all that is important has come up several times in history. This is a major misconception and was never considered a correct assessment by mainstream scientists. But maybe now we have discovered all that is worth knowing and it would be prudent to seek answers in other directions. Possibly, the final answer should lead us to consider metaphysics, philosophy, or religion for our quest to understand *why* and *how*. Perhaps if the GUT or the Theory of Everything or the ultimate particle and force are discovered it will be the end of physics as we know it. But biology, genetics, ecology, complexity, and related fields are currently approaching the stages of development that physics achieved in the eighteenth and nineteenth centuries.

Another thought is that there are limits to rational knowledge because of the physical limitations of our brains. As part of the universe, we are also regulated by the laws of physics. This philosophy believes that we will never find *the answer*.

Chapter 5

Astrology, Astronomy, and Cosmology

INTRODUCTION

There are probably more misbeliefs, misconceptions, and misunderstandings that confuse astrology with astronomy and cosmology than are found in any other area of science. To help clarify the situation, we will start with a short definition of each term, followed by a short history. We trace the development of theories and knowledge about objects in the universe, including our solar system. Where possible, we introduce the people who contributed to our understanding of these theories. It was the insight of these great observers and thinkers who provided the building blocks for our current understanding of astrology, astronomy, and cosmology.

The majority of false beliefs and misunderstandings were held by ancient people, who tried to cope with their environment by exploring or exploiting the heavens. Men and women recognized the periodicity of the movement of heavenly bodies, but without understanding it. Celestial movements and changes occurring on Earth became important to their religions as well as the flow of daily life, including agriculture.

As our understanding of the movement and nature of heavenly bodies increased, our scientific misconceptions about astronomy and cosmology decreased. As we shall see, there is still much that is unknown about our universe. Even so, we know much more than our ancestors did just a few generations ago. Also, we shall see that astronomy and cosmology are not tradi-

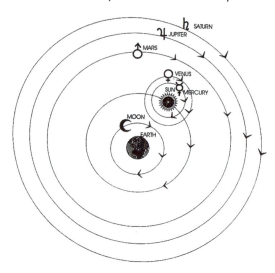

Figure 5.1 An artist's version of the geocentric (Earth-centered) solar system conceived by Ptolemy of Alexandria. The concept of perfect circles for the paths of the planets was originally proposed by Aristotle, because as he said, "God being perfect, He created a perfect universe, and thus, planets must move in perfect circles."

tional sciences for which experimental and control situations can be arranged. These areas of nature remain theoretical sciences based on observational evidence and mathematics. In fact, they are currently in danger of becoming esoteric or metaphysical sciences.

We have already mentioned some of the ancient Greeks who proposed many models for the universe—most of their concepts included a number of spheres containing the moon, sun, planets, and stars—with the Earth at the center. Some models were quite elaborate. Aristotle (384–322 B.C.) developed a beautiful, symmetrical model of the universe consisting of a geocentric grouping of spheres, but his design was wrong. Even so, it influenced many astronomers for centuries. Alcmaeon's model (about 500 B.C.) tried to improve on Aristotle's by including 55 independent spheres for the planets and stars, with air (æther) inside each sphere. Much later, in 1538 Girolamo Fracastoro designed a more elegant model of 77 nonintegrated spheres, but with the Earth still at the center. Most astronomers had a difficult time rejecting the original Aristotelian model of the universe. Ptolemy's Earth-centered model kept the sphere concept (see Figure 5.1). A few ancient astronomers had ideas about a sun-centered solar system, but these models did not match the popularity of Aristotle's original concept. It was not until the mid-sixteenth century that the Copernican and Tychonic systems were accepted (see Figures 5.2 and 5.3). This was the beginning of modern astronomy.

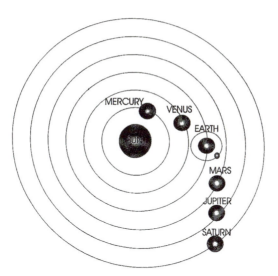

Figure 5.2 An artist's conception of Nicolaus Copernicus's heliocentric (sun-centered) solar system, indicating his belief that planets travel in perfect circles.

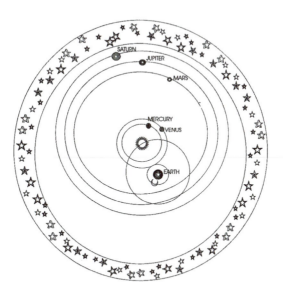

Figure 5.3 An artist's conception of Tycho Brahe's heliocentric solar system, indicating perfect circular motion of the planets and the celestial dome (sphere) of stars.

DEFINITIONS

Astrology

We should start with astrology because it has influenced men and women since the beginning of time. *Astrology* can be defined as the use of astronomical phenomena such as the positions and aspects of the stars and planets to predict earthly and human events. Thus it is a pseudoscience as there is no way to check the validity or accuracy of such predictions. Early people from all areas of the world associated the objects in the sky, and their movements, with happenings on Earth. They studied the heavenly bodies and tried to understand how they influenced human life, including their religious and personal lives as well as communal affairs. This is still the goal of modern-day soothsayers who call themselves astrologers. Even so, for many centuries, no distinction was made between astrology and astronomy. This lack of distinction still exists. Historically, the Babylonians studied and developed astrology. In later sections of this chapter examples are given of how ancient astrology contributed to people's understanding of time, agriculture, and celestial events, and how it influenced modern astronomy and cosmology, as well as mathematics.

Astronomy

Astronomy is the scientific study of celestial bodies in the universe. It grew out of ancient astrology. The main difference between ancient astrology/astronomy and the more modern science is in how knowledge is gained and used. Astronomy began about 3,000 B.C. when ancient Mesopotamians, Egyptians, and Chinese began grouping stars into constellations. Hellenistic (Greek) astronomy was more like a myth-based astrology than what we now consider astronomy. Modern astronomy began with Copernicus's (1473–1543) proposal of a sun-centered universe (solar system). Astronomy did not become a systematic science until Newton (1642–1727) introduced the laws of motion and gravity, which could be used to explain planetary and other celestial motion. It became a more exact science in the nineteenth and twentieth centuries when modern instrumentation could back-up theories with observations, and appropriated mathematics further reinforced proposed theories.

There are several areas of study in modern astronomy that require specific definitions. *Archaeoastronomy* is the study of the astrological concepts of prehistoric people based on stone structures (e.g., Stonehenge in England), cave and rock paintings, calendars, and so forth by archaeologists and astronomers. *Astrometry* is the study of the motions and positions of all celestial bodies. *Celestial mechanics* is the mathematical study of how gravity affects the motion of celestial bodies. *Astrophysics* is the study of the physical makeup of celestial bodies as analyzed by the laws of physics, and the chemical composition as analyzed by spectrum analysis of the electromagnetic energy received from these bodies. Astrophysics is also concerned with the theoretical origins of celestial bodies based on physical laws.

Cosmology

Although astronomy may be thought of as the study of individual celestial bodies, cosmology looks at the big picture of the universe. Cosmology provides much more comprehensive theories for the study and science of the universe. *Cosmology* is the study of the universe on both the smallest and largest of scales in terms of time, space, and the makeup of the universe. It includes theories about the origin of the universe and everything in it, the evolution of the universe from past—to present—to future, and the structure of the universe and its celestial bodies at various stages of their evolution. As we progress in identifying beliefs and misconceptions of specific aspects of astronomy and cosmology, there will be little distinction made between the branches of astronomy and the whole of cosmology.

Mathematics

Astronomy, cosmology, and mathematics are included together because in many ways the historical scientific development of astrology and mathematics parallel each other. Astronomy and mathematics concomitantly contribute much to the understanding of each other. We could not begin to understand our universe without the development and use of mathematics to study the universe.

There are several ways to define *mathematics.* One is the study of how many or how few, how big or small, how long or how short, or how far or how near. In other words, in the quantities, magnitudes, and relationships between objects or symbols. Another more technical definition is that mathematics is a system built on principles related to numbers and spatial relationships. Still another is that it is the body of knowledge based on specific axioms and assumptions that may or may not be proven in the real world.

Again, as with astronomy, there are branches or divisions of mathematics. There are two main divisions: *pure mathematics,* which is concerned only with the theoretical aspects of its concepts, not the practical, everyday use of arithmetic, and *applied mathematics,* which is the practical mathematics used in everyday life for commerce, engineering, construction, communicating, and so forth. Both are used in most of the other sciences. There are several other branches of mathematics: *arithmetic,* which ranges from simple numerical computations to abstract theories of numbers, is usually thought of as subtraction, addition, multiplication and division; *algebra* involves using letters or symbols to represent numerals in solving equations; *geometry* is a system concerned with points, lines, and surfaces (in two dimensions, or plane geometry), and solids (involving three dimensions, or solid geometry). The different types of geometry are based on different sets of assumptions about unproved and proved statements. Several other branches of mathematics include calculus, statistics, game and probability theory, topology, and trigonometry.

ASTRONOMY, THE SOLAR SYSTEM, AND THE UNIVERSE

We will examine some of the beliefs and misconceptions that developed from ancient to modern times regarding astronomy, our solar system, and the universe in general.

The Inner Planets

The first four planets are referred to as the *inner* or *inferior planets* because of their closeness to the sun. They are also called the "terrestrial" planets because they are close to the Earth and in some ways resemble the Earth in both the composition of rocks and minerals and size. Compared to the outer planets, they are relatively small and all, except Mercury, which has no atmosphere, have a thinner atmosphere than do the other planets. The planet closest to the sun is Mercury, followed by Venus, Earth, and Mars. Because there have been many beliefs and misconceptions about the Earth's moon, we will include it in the section on the Earth. The ancients considered the planets to be moving stars, whereas the stars were considered fixed.

Mercury

Size: Mercury is the second smallest planet (after Pluto). Diameter at equator = 3,301 miles; radius = 1,506 miles. Its diameter is about 40% of the Earth's.

Mass: 3.30×10^{26} grams; mass compared to Earth = 0.0553; mean density = 5.4 g/cc. Its mass is about 6% of the Earth's (the Earth is about 16.6 times heavier than Mercury).

Motion: Rotation period = 58.65 days, thus all surface areas are exposed to the sun; orbital revolution (sidereal) period = 88 days. It spins on its axis only three times for each two times it orbits the sun (a day is about one and a third years long).

Distances: Mean distance from sun is 36 million miles. (Because its orbit is not a circle, its distance from the sun varies from 29 to 43 million miles.)

Temperature: Daytime at the equator = 430°C to 470°C, which is about 1,380°F at one of its two "hot spots"; nighttime temperatures can drop to −180°C, and possibly as low as −200°C.

Natural Satellites: None.

Because Mercury is the innermost planet and has the smallest orbital path around the sun, it has been difficult to observe and study. Mercury's orbit is inside the Earth's orbit. Therefore it's visible path starts in one direction and then reverses. For the ancient observers, who believed in a static Earth as the center of the universe, this motion was difficult to explain. Its quick trip around the sun, only 88 Earth days, is accomplished at 30 miles per second. This rapid movement most likely influenced its being named after the Roman god, Mercury, who was named after the Greek god, Hermes, both of whom were speedy messengers.

At times Mercury has been considered more like a moon than a planet, but its composition and motion are unlike the Earth's moon. Even though it has a hot, dry, dusty, rocky, and cratered surface, it has more and larger flat plains than does the moon. Most of what we know about Mercury has been obtained by the flight of the unmanned spacecraft, Mariner 10, launched on November 3, 1973. Mariner 10 made three passes around Mercury and was able to photograph about 40% of its surface and send pictures via signals back to Earth. Because of its small mass the heat and low gravity prevented any gases from collecting on Mercury's surface. Mariner 10, and more recent radar studies, indicated that there are traces of what may be water ice at the poles, traces of some atoms of elements (potassium and sodium) that may seep up from its interior, and traces of some gases, such as oxygen, helium, and possibly hydrogen that are "blown" onto the planet by solar wind. Mariner 10 also determined that Mercury has a molten iron core with a magnetic field about 1% as strong as Earth's magnetic field.

Venus

Size: It is slightly smaller than Earth: mean diameter = 7,521 miles.

Mass: Venus's mass is 4.87×10^{24} kg (81.5% of the Earth's); its density = 90% of the Earth's.

Motions: Sidereal revolution period = 224.7 (Earth days for one orbit around the sun, i.e., a Venetian year); synodic period = 584 days; Venus's rotation on its axis is opposite (retrograde or backward) the Earth's rotation, and has a period of rotation of 243.09 days. Venus has the greatest degree of inclination of its equator to its orbital plane; it is 178°, whereas the Earth's inclination (ecliptic) is 23.4°.

Distances: Its mean distance from the sun = 67.2 million miles.

Temperature: 475°C to 480°C (900°F); this high surface temperature will melt lead and tin and will boil off any liquid water that might have been on its surface. Venus's high surface temperature is thought to be caused by the thick cloud cover that creates a "greenhouse" effect, making the surface of Venus hotter than Mercury.

Natural Satellites: None.

Because Venus is the most brilliant object in the sky next to the sun, ancient people mistakenly called it a star. They considered it the morning and evening star. Its motion across the night sky was predictable, but more erratic than the real stars. Venus's motion was recorded as far back as the first Babylonian dynasty, making it one of the first night objects to be tracked. The Babylonians called it the "mistress of the heavens" (*Nin-dar-anna*); it was used for astrological omens and to make predictions. One such prediction states that when Venus appears in the east, there will be devastations of rain from the heavens, and that when Venus appears in the west, there will be hostilities and good crops. There is some indication that the ancients knew about the phases and crescent of Venus. With good eyesight and a clear night in dry desert air, it is possible to view this phenomenon without a telescope. Not much beside the planet's motion was known because of its heavy cloud cover. Even using telescopes, astronomers could not peer through the clouds to the surface. This led to the misconception that the cloud cover would reflect the sun's heat, making the planet quite comfortable and perhaps even inhabitable. Early studies indicated that Venus's atmosphere is 98% carbon dioxide. Earth has only 0.03% carbon dioxide in its atmosphere. Although the Earth has about 78% nitrogen in its atmosphere, Venus has less than 2%. Venus has hundreds of times more of some other gases, such as the noble gases, than does the Earth. Venus would be uninhabitable by humans because it only has a trace of oxygen.

In addition to radar studies more than 22 Soviet and United States unmanned spacecraft have visited the vicinity of Venus. Much of what we now know about the planet came from the data obtained by these experiments. The atmosphere of Venus consists of several layers. The lowest is much less dense and clearer than the upper layers. It extends about 20–30 miles above the surface and contains sulfuric acid particles. The next layer is 2 miles thick and is composed of large sulfur and sulfuric acid particles. At about 32 to 35 miles, another layer consists of both liquid sulfur and sulfuric particles. And the outer layer, about 40 to 45 miles above the surface, also contains sulfuric acid particles. Beyond this layer there is a hazy, not-very dense layer of water vapor and ice.

The unmanned space probes have determined that the surface of Venus is very much like the surface of the Earth. Mountains, valleys, plains, and ancient volcanoes have been recorded. Venus also has similar geology and rock structures. The probes that landed located and analyzed limestone rocks similar in composition to those on Earth. They also located radioactive rocks, as well as volcanic and basalt type rocks. The main difference is that Venus's land masses do not move as they do on Earth.

The Earth

Size: Earth is the third planet from the sun, and the fifth largest of the nine planets. Its polar circumference = 24,859.82 miles; its equatorial circumference = 24,901.55 miles.

Mass: The Earth's mass is 3.7 million billion billion pounds, or 5.9763×10^{24} kilograms; the crust makes up 0.6% of the Earth's volume, but only 0.4% of it mass; the mantle makes up 84% of the Earth's volume, but only 67% of its mass; the core makes up almost 15% of the Earth's volume, but 32% of its mass.

Total Volume: The Earth's total volume is 259.8 billion cubic miles; its water volume = 330 million cubic miles; its crust volume = 2 billion cubic miles (the crust ranges from 3 to 7 miles in thickness); its mantle volume = 216 billion cubic miles (the mantle is approximately 1,800 miles thick); its core volume = 41 billion cubic miles (the core is about 1,400 miles thick).

Densities: The density of the Earth = 5.518 g/cc; water (the standard for densities) = 1.00 g/cc; crust = 2.85 g/cc; mantle = 4.53 g/cc; core = 10.70 g/cc.

Motions: The Earth's rotation speed (on axis at equator) = 1,000 mph; its orbital speed (path around the sun) = 66,672 mph; its escape velocity (velocity required to leave Earth's gravity) = 6.96 miles per second. One rotation on its axis equals 1 day (24 hours); one revolution around the sun takes 1 year (365.4 days). The Earth is closest to the sun in January.

Temperature: The highest recorded temperature = 140°F at Delta, Mexico (highest in the US = 134°F in Death Valley, CA); the lowest recorded temperature = –128.6°F at Vostok Station in Antarctica (lowest in US = –79.8°F in Prospect Creek, Alaska); its average global temperature = 59.8°F.

Natural Satellites: One, Luna, our moon.

The myths associated with the formation of both the universe (world) and life have been mentioned. Creation myths also include the so-called "cosmic egg," used as a metaphor for the emergence of humans, as well as the Earth, brought up from primordial waters.

Prehistoric men and women had many misconceptions about the Earth that grew out of their desire to understand their environment. They surely knew that if a stone was thrown into the air it would return to Earth. The reason for this attraction to the Earth was not understood for many thousands of years, yet people learned how to work around this peculiarity. Early people mistakenly believed that the Earth was flat. After all, as far as the eye could see there were trees, hills, plains, or water. It is still believed by some people that the Earth is flat.

A misconception about the Earth that persisted for several thousands of years was that the sun, moon, and heavens did the turning—not the Earth. The rationale for this mistaken belief was Aristotle's statement that earth and water move downward and air and fire upward, and that all heavenly bodies moved in circles. This misconception was believed and taught into the eighteenth century, even after proven false. The great astronomer, Ptolemy, said that a moving (rotating) Earth would be contrary to nature because everything would be torn loose from the Earth and thrown off into space. He insisted that if you dropped something on a spinning Earth, that it would land west of where you dropped it because the Earth would have moved east. Everything, including clouds, would be moving westward if the Earth rotated eastward. It has been speculated as to why Ptolemy did not consider the same problems with the spinning of the celestial sphere, since according to him it would be spinning at a much greater rate then would a spinning Earth.

It is not known when people recognized the periodicity of the movements of the Earth, moon, sun, and stars in the celestial sphere. Early in history it was soon learned that the reoccurrence of the sun each morning, the moon each month, and the seasons each year could not only be counted on but measured. The lack of understanding of what caused these regular events led to numerous misconceptions. For example, an ancient myth proclaimed that some demon ate the sun each day, and it was reborn the next morning. Or that the seasons were caused by gods, for whom they were named. And that the lunar months were related to fertility, both in harvests and human reproduction, based on the menstrual cycle of human females. But at the same time, observations led to the development of simple arithmetic and counting systems, which in turn led to the development of mathematics and calendars. A major problem that complicated all early heavenly observations is that the Earth has several different motions that do not factor as whole numbers (e.g., the number of days does not divide evenly into the length of the year, nor do the phases of the moon easily equate with the length of the year). Over the centuries these multiple motions became even more complicated when it was decided that the Earth did not, and could not, move because it was the center of everything, including the spiritual center of the ancient world (universe). These led to assumptions that all other celestial bodies did the moving, which they do, but not as observed and recorded by ancient astronomers.

The crude measurements, records, and calendars guided people in ways to control some of their activities, such as moving to new hunting grounds and gathering plants based on the seasons. Later, when agriculture was established, record keeping of celestial movements was essential to community welfare. Both observations and computations were required for astronomers (astrologers) to make forecasts and predictions. The numbers were read off crude instruments, but astronomical quantities were derived by computations—thus mathematics grew up with astronomy—each dependent on the advancement of the other. Tally sticks, nicks cut into sticks, knots on strings, stones with marks on them indicate that early people used systems to keep track of things. An early example is a prehistoric bone of a wolf with 55 notches carved on it. The marks are arranged in groups of five, followed by a larger mark, and then repeated. It is speculated that this counting system was based on the five fingers of each hand. We still make tallies of four vertical slashes with a single cross slash to record things in fives. There is some evidence that body parts were used as standards for measurements; for example, an arm's length. Ancient people surely had sounds or signs to represent the concepts of few or many, small or large, or more or less. They no doubt used their fingers (and possibly toes) and other means of determining quantities (amounts) as well as qualities (size, color, etc.) of objects. For example, each day that the sun shone there was a particular time when it was highest in the sky (zenith), but not always directly overhead. From this it would be a simple exercise to mark off the sequence of hours and days as the sun reached its zenith. There was a problem with this approach, however, because humans will go blind if they look directly for too long at the sun. They solved this problem by driving a stick into the ground or erecting a stone pillar to cast a shadow. The shadow would be longer at dawn, shorter at noon, and longer again at dusk. It wasn't long before a device was constructed to mark off the lengths of the shadows as the sun progressed east to west. Such a sundial system was used in Egypt around 3,000 B.C., but it could only measure shadow segments during the days when the sun was shining. There were many misconceptions as to why the sun would repeat this performance. One was that at sundown the sun was burned out and had to go away, possibly to an underworld, to come back the next morning, reborn and ready to repeat its cycle. In general, the ancients knew that the sun was "hot" but could not understand why it did not burn out as a wood fire did, so they had to develop explanations as to how it could be regenerated each day.

Telling the approximate time and even seasons could be accomplished by identifying different star patterns called constellations. Recognized constellations would appear at different times of the year in the eastern sky just after sundown and move westward. Once a record was made of the movements, timekeeping at night and the seasons could be estimated. The water clock was invented as a timekeeping device in about 1,400 B.C. Candles with marks on them were used sometime later. These were the best nighttime timekeeping devices that were available for many centuries. The moon was also observed and used to track time, but on a monthly basis, rather than hourly or daily.

Early people also mistakenly believed that the Earth was much smaller and less massive than it actually is, and that it rested on a great body of water. They believed the Earth and the heavens were controlled by spirits and/or demons. These beliefs led to the era of mythology, which again existed simultaneously in history, in several different countries. Because the Earth provided the sustenance for life, the Greeks considered it a living organism (mother Earth) they called Gaia.

As people learned more about the Earth's makeup and motions, a number of beliefs promulgated superstitions and misconceptions about the Earth. We will consider a few examples.

A retired army captain, John Cleves Symmes, spent his days trying to convince everyone that the Earth was composed of five concentric spheres with large openings at the poles. In 1818 he tried to organize an expedition to the north pole "opening" to explore "Symmes's hole." He believed that the oceans flowed through the holes and supported life on the inside of other spheres. He even approached the U.S. Congress for support, where he received 25 votes in favor of his plan. His misconception was based on a belief held by Cotton Mather (1663–1728), the Boston minister of the Salem witchcraft trials, who published his "hollow earth" theory in 1721. Mather's theory was based on misconceptions held by Edmund Halley (1656–1742), the English astronomer and discoverer of Halley's comet. Halley believed that the Earth had a 500-mile-thick shell that contained two smaller spheres the size of Venus and Mars, with a core the size of Mercury. He also believed that life existed in these inner Earth spheres.

In the late 1800s Cyrus Reed Teed spent years lecturing and writing about his misconceptions. He believed that the Earth was completely hollow, with an outer shell only 100 miles thick, and that life existed on the inside, not on the outer surface. This 100-mile-thick shell had 17 layers, some form the geological layer, others the metal and mineral layers. He was devoutly religious and could not accept the concept of infinite space, so he devised his "womb" theory based on Holy Scriptures. He took the name "Koresh" about the time he published his revelations. Teed was also an alchemist and an herbalist who had a vision of a beautiful women who gave him the idea that the entire cosmos is like an egg and that we live on the inside of this egg, in what we call the Earth. Inside the hollow Earth are the sun, moon, planets, and the stars. Outside the Earth is just nothing. His interior sun is half light and half dark, thus forming a reflection that we see as the illusion of a rising and setting sun. He developed more and more elaborate ideas to back up his original misconceptions.

Later in 1913, Marshall B. Gardner published his theory, which he insisted was not based on Symmes's ideas. His misconception was that the Earth's outer shell was 800 miles thick, and the entire interior was hollow, except for a 600-mile sun that maintained continual daylight for the interior of the Earth. He also believed that there were openings at both poles and that all the other planets were constructed in the same manner. His proof was that the ice caps on the poles of Mars were the openings into the center of Mars, and that the aurora borealis (Northern lights) seen at the northern regions of the Earth is caused by light escaping from the Earth's northern opening.

The 100-year span from 1850 to 1950 saw many misconceptions as to the origin of the Earth and other planets. We have already mentioned Hans Horbiger, an amateur astronomer who formed a mystical, anti-intellectual cult of millions known as WEL (German for the cosmic ice theory)—it still exists as a viable cult. Their main misconception is that the moon and planets are covered with many miles of ice.

Ignatius Donnelly, was a U.S. congressman and later a state senator. He believed in the lost continent of Atlantis; that there are astrological "cipher messages" in Shakespeare's writing; and that a former comet wrought catastrophic effects on the Earth. (For an update on comet theory see the section, Other Objects in the Universe, in this chapter)

William Whiston, who succeeded Sir Isaac Newton as professor of mathematics at Cambridge, published his *New Theory of the Earth* in 1696. Even though most religions at that time accepted the concept of a spherical Earth as one of the planets orbiting the sun, Whiston believed otherwise. His misconceptions were many. He believed that the "void of chaos" was the tail of a giant comet. He believed that the Earth, planets, and moons were formed from this comet's tail. He believed that all bodies revolved in perfect circles. He also believed that the year consisted of exactly 360 days, and that originally the Earth did not spin (rotate) on its axis; therefore an original day was also the same as his 360-day year. Of course it was known, even then, that there are 365.25 days in a year. Whiston also believed that the moon made a complete circle of the Earth in exactly 30 days. He also stated that after Adam and Eve ate the forbidden fruit, the comet's tail caused the Earth to start spinning on its axis. He dated the arrival of a second comet on Friday, November 28th, in the year 2349 B.C., which was sent to visit Earth to show God's displeasure with the wicked world. It caused 40 days and nights of rain that condensed from the tail of the comet. Only Noah knew the exact date of its arrival, so he was able to build his ark and save the animals. This second comet also caused the Earth to start revolving around the sun in an elliptical path instead of in a perfect circle. Whiston claimed that the inertia of all this water, plus magnetic forces, increased the length of the year to 365 days. He believed that all this excess water drained into the center of the

Earth. He backed up his theories with mathematics and religious texts. The scientific climate of this period was such that several other scientists considered him to be a learned colleague who proposed a viable account of catastrophes caused by comets.

Immanuel Velikovsky's *Worlds in Collision* (1950) is mostly based on legends, folktales, and myths from old manuscripts and biblical texts that many believed relate to the catastrophes of ancient times. Many people of his day ranked Velikovsky with Galileo, Newton, Kepler, Darwin, and even Einstein, even though evidence for his theories was never produced. Several science fiction novels and movies resulted from his book and misconceptions.

There were many old-time superstitions about the surface of the Earth—or rather misconceptions people held about the magical power of just plain dirt, sometimes referred to as "fruitful dust." Dirt, particularly dirt from a churchyard, was thought to have special healing power that would even ease the pain of birth and death. Dirt was also used to heal infected toenails, breast cancer, rheumatism, influenza, warts, and, when placed under a hat, would protect against witchcraft. Likewise, if this holy dirt was placed in a person's coffin, his or her soul would be able to enter the next world before many moons had passed.

Moon

Size: The moon's equatorial diameter = 2,160 miles; about one-fourth of the Earth's size; our moon is the fifth largest of all the planetary moons in the solar system.

Mass: The moon's mass is 0.073×10^{27} grams; its density is 3.34 g/cm^3.

Motion: The moon's rotation on it axis and its period of revolution around the Earth are the same: 27.32 days. Ancient tidal effects and the pull of gravity between the moon and Earth caused a "despinning" of the moon until it obtained a stable synchronous period, which resulted in the same side of the moon always facing the Earth. The moon "rises" about 52 minutes later each night.

Distances: The moon's average distance from Earth is 238,865 miles; at perigee (closest) the distance = 222,755 miles; at apogee (furthest) the distance = 254,185 miles.

Temperature: The sunlit side of the moon is 273°F; the dark side, –243°F.

There were many ancient beliefs and misconceptions about the moon. Some persisted up to the twentieth century. All cultures connected the moon with religions, festivals, and often held orgies under the full moon. The ancient Eskimos held feasts, darkened their lamps, and then exchanged women. Ancient Africans chanted to the moon and held dances. Ancient

Germans held their business meetings at either the new or full moon because they believed these were the best times for commerce.

The English word for *moon* was derived from the Greek word "meter," which means "to measure." This relates to the moon being the first instrument used by man to measure time longer than a day. One of its earliest uses was to keep track of the female menstrual cycle. A pregnant woman could expect a child after missing her menses for 10 moon months. It soon proved inaccurate for predicting the best times to plant and hunt, because the moon's periods do not coincide equally with the seasons. Over a period of several years a "moon calendar" became so inaccurate that it could no longer be used to predict the first frost, the rainy season, or the correct dates of religious significance. It required some time and study to realize that the seasons and the length of the year were determined by the orbiting of the Earth around the sun, and not the moon orbiting around the Earth. Some civilizations, such as the Babylonians, suffered by maintaining a lunar calendar, whereas others, such as the Egyptians, recognized the problem and used different methods to determine the length of a year.

The moon, being the closest heavenly body to the Earth, was visible during its phases with the unaided eye. One misconception was that the moon, like all wandering objects in the sky, was made of crystal and its surface was perfect and really smooth. It disturbed people when imperfections on its surface could be seen. One explanation was that the Earth soiled the surface of the moon, making it imperfect. The moon was the only object in the sky that visibly changed from night to night. Its steady and repetitive changes were much studied. Speculations about its composition and surface were many. Some insisted they could see a face on the moon, depicted as the "man in the moon."

There were also speculations as to the possibility of life on the moon, or at least that its changes resembled a living thing. Even the great astronomer, Johannes Kepler (1571–1630), who supported Galileo's telescopic observations of the moon and Jupiter, believed that there was life on the moon. In 1835 Richard A. Locke wrote articles for a New York newspaper about his discovery of advanced life on the moon. He knew it was fiction, but the public bought it, and a lot of newspapers were sold.

The ancient Greek explorer Pytheas sailed out of the Mediterranean sea, into the Atlantic, to the British Isles in about the year 300 B.C. He experienced coastal tides for the first time. He thought that the moon had something to do with the tides because of their regularity. When Galileo heard of this in the early seventeenth century, he insisted that the moon had absolutely no affects on the Earth whatsoever. One of Galileo's misconceptions was that he believed the tides were caused by the rotation of the Earth, which, in turn, caused the sloshing of the ocean water onto the beaches. The problem with Galileo's reasoning was that this sloshing should only happen once a day instead of at least twice a day. The final tidal explanation had to wait for Newton's concept of gravity. Tides are produced by the gravitational forces of the moon and sun, which pull the Earth out of shape about 1 foot. (The sun's gravitational attraction of the Earth is about 40% that of the moon because the sun is much farther from the Earth than it is from the moon.) At the time the gravity of the moon is attracting the solid Earth, it also pulls water toward the moon, causing high tide. At the same time, on the other side of the Earth, the water (and to a lesser extent the Earth itself) is pulled to the side opposite the one near the moon—thus creating a low tide on the side of the Earth opposite the moon. During a full moon, when the moon and the sun are in a straight line on opposite sides of the Earth, their combined gravitational pull creates an extra high tide called a *spring tide.* When the moon and sun are at a 90° right angle to each other, an extra low tide called a *neap tide* is produced by the counteracting gravities of the moon and sun.

At one time, astronomers held the misconception that the craters on the moon were caused by volcanoes. Later, it was demonstrated that volcanoes on Earth always appeared on a mountain-like cone, and the craters on the moon's surface were not raised as cones. The reason is that when a meteor strikes the moon's surface, it explodes and digs a hole forming the crater, throwing out dust that forms rays on the surface. Because there is little or no atmosphere, water, or wind, there is no erosion to wear down and fill in the craters. This is true for all airless, waterless bodies in space.

There are a number of beliefs and theories as to the origin of the moon. One proposed by George Darwin, the son of Charles Darwin, is called the *fission theory* in which the ancient, rapidly spinning Earth flung off a huge chunk to form the moon. Supposedly, this chunk came from where the Pacific ocean now exists. The *capture theory* states that the moon, somewhat like a small planet or large meteor, was captured by the Earth's gravity as it came flying by in ancient times. Another theory was that space dust condensed to form the planets and their satellites, including the moon. Yet another, more acceptable idea is the *cataclysmic theory,* which states that a large body (Mars' size) struck the ancient primordial Earth and threw debris into space that coalesced to form the moon. In time, most of the debris in space was attracted to planets, as smaller things were attracted to larger bodies. This could explain the so-called *double-planet* concept of the Earth–moon pair. Another, more recent theory is that a giant meteor struck the Earth in the area of the Yucatan peninsula in Mexico. This resulted in a huge chunk of Earth being sent into space to form the moon. The question still awaits a final answer.

Between 1968 and 1972 there were nine manned space flights to the moon, and 12 American astronauts walked on the moon. Hundreds of pounds of moon rocks and soil dust have been returned to Earth by the astronauts. Most are now in storage but some rocks are available for study; others are exhibited for the public to view and touch. The rocks are similar to Earth's volcanic rocks. They have a higher content of titanium, and a lower content of iron than do Earth rocks. The moon's surface is covered with moondust, rocks, as well as domes, ridges, mountains, flat areas called *mares* (moon "seas"), and many craters with rays spreading out from their centers. These lighter colored rays are evidence of the impacts of meteors on the surface of the moon. There is some evidence of past volcanic activity. Instruments left behind by the astronauts send signals indicating that the moon has tremors and "moonquakes." Like the Earth, the moon has a thick crust, a mantle, and possibly a molten core.

Mars

Size: Mars is the fourth planet from the sun, and is about half the size of the Earth and twice the size of the moon. The diameter at its equator = 4,220 miles; its diameter at the poles = 4,195 miles.

Mass: Its total mass is about one-tenth that of the Earth or 6.42×1023 kg, which gives it a gravity of about 38% as strong as Earth's.

Density: Its density is lower than Earth's, 3.9 g/cm^3, compared with 5.5 g/cm^3 for the Earth.

Motion: Because Mars is farther from the sun than the Earth, it takes longer to complete 1 year (one revolution around the sun). Its orbital speed around the sun is 15 miles per second. Its rotational period (a day) is 24 hours, 37 minutes, and 22.3 seconds long (Earth time); its year is 687 days.

Distances: Its elliptical orbit is about one-and-a-half times as far from the sun as the Earth's. The mean distance from the sun = 141.5 million miles. Its perihelion (minimum distance) is 128.4 miles from the sun. Its aphelion (greatest distance) is 154.8 miles from the sun. Its inclination from its planetary plane is 25.2°, compared to Earth's 23.4°.

Temperature: During the polar night the temperature varies from –199°F to –64°F. When Mars is closest to the sun, the daytime temperature is 80°F.

Natural Satellites: Two: Phobos (Greek for "fear") and Deimos (Greek for "terror"), both named after the two sons of Ares, the Greek god of war, who in Roman mythology became Mars. These moons are relatively small rocks with many craters, similar to asteroids.

The planet Mars has intrigued people for thousands of years. Beliefs and misconceptions about its nature and the possibility of life existing there have been recorded more often than for all the other planets combined

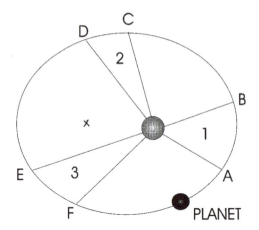

Figure 5.4 Kepler's first law of motion states that planets follow an elliptical orbit around the sun. Sections 1, 2, and 3 are equal areas (not to scale).

(except Earth). Because of Mars' reddish color, it was named after the Roman god for war.

Mars makes a looping path in the sky because of its elliptical planetary orbit, which lies just outside Earth's path around the sun. This caused much confusion until it was decided that the sun, not the Earth, was the center of the solar system. Johannes Kepler (1571–1630) determined that the path of Mars was not circular as Plato, Aristotle, Ptolemy, and others mistakenly believed, but rather its orbit was elliptical. Kepler's first law states that the sun is just one of the two focal points in the elliptical path of planets. The sun is the main focal point, whereas the second focal point is imaginary. Kepler's famous second law states that an imaginary straight line joining a planet to the sun sweeps out equal areas of the ellipse in space in equal intervals of time. This means that the time a planet travels for a given segment of its orbit that is furthest from the sun, equals the same time it travels in its orbit when closest to the sun, *if* the areas of the triangles formed between the segment of the orbital paths extended to the sun are equal. In other words, planets travel faster (but cover a greater distances) in their elliptical orbits when closest to the sun, and they slow down as they travel the shorter distances in their elliptical orbits, when they are furthest from the sun (see Figures 5.4 and 5.5).

Galileo Galilei (1564–1642) made the first telescopic observations of Mars, as well as the moon, Jupiter, Venus, and other objects in the sky. He was able to observe some of Mars' surface features, but not clearly. Galileo believed planets moved in perfect circles because things should be organized, simple, and perfect. In the late 1800s telescopes improved enough to enable more details to become visible on the surface of Mars. Giovanni Schiaparelli

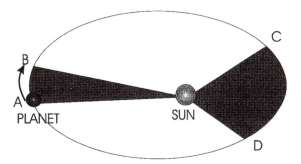

Figure 5.5 Kepler's second law states that as a planet moves in its elliptical path around the sun, it sweeps through equal areas in equal times. The two shaded sections denote equal areas. Thus a planet travelling the arc between **C** and **D** must travel faster than when moving along the arc between **A** and **B**.

observed what he considered "canali" or canals that seemed to connect areas of the Martian surface. Many astronomers believed this was proof of life on Mars. This misconception was promoted by a well-known astronomer from the United States, Percival Lowell (1855–1916).

Lowell was convinced that intelligent life existed on Mars. To promote his beliefs he wrote a book entitled *Mars*, which included many of his photographs of Mars and maps he drew depicting over 500 canals. Other astronomers disputed the existence of the so-called canals on Mars that Lowell and Schiaparelli reported. Later experiments were conducted that explained that the canals were most likely flaws or aberrations in the lenses of the telescopes, or optical illusions caused by connecting dark areas on the surface of Mars together with lines, or they were the product of overactive imaginations. To his credit, Lowell, a rich man, built an important observatory in Arizona and made many observations of the movements of Uranus and Neptune. From these observations he anticipated the discovery of Pluto.

Over 13 unmanned spacecraft have either flown by, orbited, or landed on Mars. Not all of the missions were successful, but the amount of data gathered is tremendous. We have learned much more about Mars in the 25 years of space flight than in all past history. Mars's atmosphere is 95.3% carbon dioxide, 2.7% nitrogen, and 1.6% argon. There are only traces of oxygen, water vapor, and other gases. This atmosphere is so thin that its surface pressure (about 1 psi) is less than one-tenth of the Earth's. There is a thin atmosphere composed mostly of carbon dioxide. There are air currents (wind), clouds, fog, and dust storms. Frozen polar caps of solid carbon dioxide, and possibly some water, advance and recede with the season. The daytime temperature is a mild 80°F. Mars has a gravity that is less than Earth's. No evidence of any simple or complex life has yet been found on Mars itself, and probably never will be. It is an inhospitable planet composed of mostly old basalt type rocks, iron oxide dust that makes it look reddish, pebbles, small rocks, and sand. It has no rivers, lakes, or oceans so the only current erosion is caused by wind-blown dust. There is evidence of canyons, plateaus, and some volcanic activity. One small volcano, Elysium, is 3 miles high. Conditions on Mars make it impossible for earthly forms of life to survive unless assisted by artificial means.

Some of what we know about the composition of Mars comes from 12 meteorites that were knocked off its surface and landed on Earth. A few were found in the ice at Antarctica. One of the Martian rocks that landed on Antarctica is of special interest. This Mars rock contains fossilized evidence of simple organic molecules, which indicates that some form of simple life may have existed on Mars in the past. The possibility of former life on Mars is debated by scientists.

July 4, 1997 was the landing day on Mars for an unmanned spacecraft launched in December of the previous year. It was designed to gather geological information about the red planet and quell qualms about possible life on Mars. It sent back to Earth some amazing color and 3-D photographs, as well as valuable geological information. In addition to striking pictures of the surrounding areas, the "rover" analyzed rocks and the dusty surface. The soil samples indicate similar soils to those analyzed by the Viking landers over 21 years earlier. The first rock it analyzed, "Barnacle Bill," contained some quartz, indicating that the material had been melted and remelted several times, which indicates ancient volcanic activity. There is unmistakable evidence of giant flood planes and rock erosion, and thus a history of water on Mars. An early misconception made by the media involved statements that the descent of the parachute apparatus fell to the surface of Mars ten times faster than it would have on Earth because Mars has greater gravity. They must have been thinking of the difference in the density and pressure of the atmospheres between the two planets. Because the atmosphere on Mars is only about one-tenth that of Earth's, a parachute descending on Mars would drop faster than it would on Earth because Earth's atmosphere is much *denser. Martian gravity is less than 40% that of the Earth, not ten times greater.*

The large gap in the heavens between Mars and the next planet, Jupiter, was thought to contain a lost planet. Objects called *asteroids* were later discovered in this region. They will be included in the section about other bodies in the solar system.

The Outer Planets

There are five outer planets, sometimes referred to as the *superior* or *Jovian planets.* They are not only a greater distance from the sun than are the inner planets—some a very much greater distance—but most of them are much larger than the inner planets and have different physical characteristics. They are composed mostly of gases, with a possible core of ice–rock. They are, in order of the distance beyond Mars and from the sun, Jupiter, Saturn, Uranus, Neptune, and Pluto, followed by a yet to be discovered Planet X. The outer planets are composed of refractory rocks that have a very high boiling point as well as several volatile gases. The larger outer planets were formed so far from the sun that they did not lose all of their lighter gases to the sun's heat as did the inner planets. Also, their greater gravity retains lighter gases, such as, hydrogen, ammonia, methane, and some water vapor. This partially explains why they are the largest, most massive planets. There are also numerous misconceptions about the first few outer planets that were known to ancient astronomers, but the other, more distant planets were not discovered until modern times.

Jupiter

Size: Jupiter's equatorial diameter = 88,800 miles. Its polar diameter = 82,952 miles. It is the largest of all planets.

Mass: Compared to Earth's mass, Jupiter's equals 317.83, which means it is almost 318 times the mass of the Earth. It is the most massive of all planets, and is about two-thirds the *total* mass of all the other planets combined.

Density: Jupiter's density = 1.33 g/cm^3 (which is low). Its volume is about 1,000 times the Earth's; however, 1,000 Jupiters could fit inside the sun.

Motion: It makes one rotation on its axis every 9 hours, 55 minutes, and 41 seconds (Earth time), which is a rapid rotation period. It revolves in its orbit around the sun in 11.9 Earth years. Its inclination to its orbital plane is 3.1° (Earth's = 23.4°).

Distances: It is 5.2 AU (Astronomical Units) that is the average distance of the Earth from the sun. Therefore, 1 AU = 93,000,000 miles. In other words, Jupiter is 483,600,000 miles from the sun, or 5.2 times the Earth's distance from the sun (5.2 × 93,000,000).

Temperature: The temperatures vary in its multiple cloud layers. In general, the upper clouds are colder, about −130°C, and contain solid ammonia. Proceeding toward the surface, it is about −40°C and contains layers of ammonium hydrosulphide and possibly water vapor. At the surface it may be as warm as 20°C.

Natural Satellites: Jupiter has 16 natural satellites. The four discovered by Galileo, which can be seen with good binoculars, are: Io = 2,256 miles in diameter; Europa = 1,950 miles in diameter; Ganymede = 3,270 miles in diameter; and Callisto = 2,983 miles in diameter.

Jupiter is the fifth planet from the sun. The planet is named for the chief Roman god.

Jupiter, as well as Saturn and the other "wandering" stars (planets), were the main elements of astrology in Greece, Babylonia, India, China, and Egypt, as were the twelve zodiac houses of fixed stars. The seven known planets (which at that time included the sun and the moon) became of great mystical importance.

The more recent Babylonians, known as Chaldeans, were interested in Jupiter for religious purposes. Their main misconception, which lasted several centuries, was that Jupiter was a brilliant, moving star somewhat similar to Sirius, even though Sirius did not wander in the sky as did Jupiter. It was less bright than Venus, but they thought Venus was haughty or vain because of her brightness, so they gave Jupiter higher godlike status. A major misconception the Romans had about their chief god was that, as the sky god" Jupiter (Jove) was responsible for all the weather on Earth, especially lightning and rain. They also believed that Jupiter was the champion of Romans, and after victory in battle they made sacrifices to him.

All of the ancient astronomers/astrologers had trouble with their calculations and tables because they held the false belief that the heavens revolved around a stationary Earth. Even so, they developed tables that involved the wandering stars (planets), and their interactions with the zodiac constellations. These tables were used by astrologers to make more accurate predictions of how the stars affect events on Earth. By using star and planet tables based on observations of the positions and motions of heavenly bodies, astrologers, including those who practice this nonscience today, consider their readings to be more rational and thus more scientific. In the past, astronomy and astrology were somewhat synonymous and their predictions were used for political, economic, and religious purposes. Today, astronomers do not consider astrology a science because astrological readings and predictions lack validity and reliability.

Jupiter has a more regular motion in the night sky than do the other wandering stars because it makes only one retrograde loop (meaning that it appears to earthly observers to loop back on its path only once, rather than several times, as do some of the other planets). It also visits all the constellations in the zodiac in a 12-year cycle. For these reasons and because, at its closest approach to Earth, it is almost as brilliant as Venus and thus easy to observe, the ancient mythical Babylonian king, Marduk, considered Jupiter the ideal star to regulate everything else in the heavens. He appointed Jupiter the night sun and made it the supervisor of the night sky.

Other cultures also had false beliefs and misconceptions about Jupiter. In ancient China, Jupiter was known as *sui hsing* or the Year Star because it stayed 1 year in each of the twelve star zones, which led to the 12-year cycle used in their system. In ancient India, Jupiter was called *Brhaspati* for one of the Hindu gods. The ancient Europeans gave each of the two then-known outer planets *talismans* or symbolic astrological characters. It might be noted that many of these characteristics had little to do with realities related to the planets, but were mostly based on old mythologies and misconceptions. Some examples of talisman symbols are given in Table 5.1.

Table 5.1
TALISMAN SYMBOLS

Planet	Metal	Color	Stone	Animal
Jupiter	tin	blue	sapphire	eagle
Saturn	lead	black	onyx	crocodile

It was not until the early part of the seventeenth century that Galileo observed Jupiter with telescopes he built himself. Galileo's telescope had concave-shaped ocular lenses that limited it to only 30 power (about the same as modern binoculars). Later telescopes used a convex ocular lens that enabled larger lenses and telescopes to be constructed, increasing their viewing power (see Figure 5.6). Later reflecting telescopes, using mirrors to gather light, were developed that were even more powerful. Galileo wrote books on what he saw on the moon, and he reported that the planets looked much different than the stars when viewed with the aid of a telescope. The planets appeared as pale disks, whereas the stars continued to look like points of twinkling lights, only brighter. This was one of the first clues as to the actual nature of planets, and that they were not stars. He could see that the Milky Way was made up of millions of stars, and the Pleiades had over 40 stars in its constellation. One of his most important discoveries was the three smaller "planets" he saw revolving around Jupiter (later he discovered a fourth). He called them new planets and gave them the name *Medicean stars* after the Grand Duke of Tuscany. Because their nightly positions changed, there was no question that these new "planets" orbited Jupiter. This proved that the Earth was not the center of all heavenly movement. The significance is that observing the orbiting of Jupiter's moons provided the first real proof that the Earth was not the center of the universe, and it confirmed the Copernican heliocentric system.

Somewhat amazing was the reaction of many learned scholars. They rejected the concept of the new wandering stars around Jupiter because Aristotle never mentioned them. When Galileo offered them the chance to view the satellites of Jupiter, some refused. Others looked, but said they saw nothing. They said it was all illusion. Today we have a saying for this: "Don't confuse me with facts, my mind is made up!"

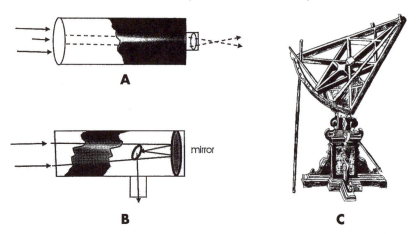

Figure 5.6 A: The refracting telescope was the type first developed and used for early observations of the planets and stars. There are limits to the sizes of lenses that can be made for this instrument; thus there are limits to its use for astronomical observations. **B:** The reflecting telescope can be constructed with a much larger mirror to collect starlight. It is the basic type of instrument used today. **C:** The large sextant constructed for Tycho Brahe is an example of some of the very large instruments developed to study the heavens.

Once astronomers could more clearly see the surfaces of the sun, moon, Jupiter, and other planets, speculation abounded as to their compositions, their atmospheres, life forms, and so on. All types of life forms were proposed for the other heavenly bodies. Kepler, Huygens, Fontenelle, and others wrote books and speculated on the types of life existing on Jupiter, the moon, and the other planets. Being religious, one thing they seemed to agree on was that living beings on other worlds could not be descendants of Adam and Eve.

Galileo saw moving clouds on the surface of Jupiter. Huygens made a drawing of Jupiter indicating two bands or streaks at the equator of Jupiter, and one dark band on Mars. Dark spots could be seen and tracked to determine the period of rotation. Cassini, using an improved telescope, determined the rotation period of Jupiter as 9 hours and 56 minutes. He did this by observing irregularities in the equatorial bands identified by Huygens. This was proof that planets were spinning on their axes. So, the Earth too must also be rotating and moving. Thus, it cannot be the center of the universe. This provided further evidence to support Copernicus's heliocentric theory, as well as his theory that mutual attraction of particles that make up spinning masses causes the spherical shape of planets.

The seventeenth century was the age of experimental and mechanical philosophers, culminating in the work of Sir Isaac Newton (1642–1727). The Earth, the solar system, and the universe were now seen as a giant clock-driven machine. Although there still was little distinction between the sci-entist and philosopher, a new way of thinking and experimenting now existed. No longer could one make deductions from *a priori* principles or ancient myths and misconceptions. From this period through modern-day astronomy and science, improved instruments and calculations were used to gather more accurate data about Jupiter and other planets.

The United States has sent six unmanned satellites to Jupiter, plus two fly-by satellites to view Jupiter on their way to examine Saturn. A great deal of information was collected about Jupiter in these flights. For example, Jupiter is composed of gases, mostly hot liquid hydrogen, which may become a solid metal at the center of the planet. There is also evidence of some helium gas and other elements mixed with the mostly liquid hydrogen. There are traces of hydrogen compounds and some organic (carbon) compounds. The center of Jupiter is about 54,000°C, and gets cooler as you proceed outward. For instance, just outside the center it is only 20,000°C, then proceeding further, it is 15,000°C, then 10,000°C, and finally at 600 feet below the surface it is only 3,600°C. The thin outer layers of clouds are several hundred degrees below freezing. It was also discovered that Jupiter's magnetic field is 40 times stronger than Earth's. This field was measured by a fly-by satellite when it was about 4 million miles from the planet. One interesting speculation, which may or may not be a misconception, is that Jupiter very nearly became a star similar to the sun. If it had collected more mass at the primordial beginning when gases and particles coalesced by gravity, fusion of the hydrogen into helium would have taken place. This is the same process that fuels the stars.

In 1994 a comet or asteroid fragment slammed into Jupiter creating gigantic explosions observed by astronomers on Earth using powerful tele-scopes. The winds of Jupiter move both east and west, depending on their latitudes. They have been clocked at many hundreds of miles per hour and are somehow related to the colored bands observed from Earth. Also, there is the mysterious, Earth-sized Giant Red Spot that seems to just keep rolling in position between wind and color zones. Recently discovered are several diffuse rings around Jupiter in the elliptical plane area of its satellite Io. These are similar to Saturn's rings, but not as extensive or prominent. The ring's particles are small enough to be affected by electromagnetic forces that drive them toward Jupiter. It is speculated that the rings are maintained by micrometeorites from space and volcanic particles from the volcanoes of Io. Many of Jupiter's satellites are small, similar to asteroids or even meteors captured by Jupiter's great gravity.

Saturn

Size: Saturn's equatorial diameter is 74,849 miles. Its polar diameter is 66,610 miles.

Mass: Saturn's mass = 95.15 the mass of Earth. Its density = 0.69 g/cm³ (which is very low). Saturn's gravity is 1.15 times that of Earth's gravity.

Motion: Its rotation period = 10 hours and 40 minutes. Its period of revolution around the sun = 29.458 Earth years. Its orbital inclination to its elliptic = 2.5°.

Distances: Saturn's average distance from the sun is 887 million miles or 9.54 astronomical units (AU).

Temperature: The internal temperatures must be high because Saturn radiates from its surface about three times the heat it receives from the sun. The outer surface is cold enough to keep all gases frozen.

Satellites: Saturn has at least 20 known natural satellites in addition to its rings.

The major misconception ancient people had about Saturn was that it was a wandering star. They made this same mistake about the other planets. Saturn travels very slowly across the heavens when compared with Mercury. This led to the concept of Saturn being Father Time. To the Romans it was known as *Saturnalia,* or their old harvest god, *Saturnus.*

Saturn, like the other planets, would proceed across the sky, then occasionally stop and move backward (retrograde motion), which confused ancient observers for centuries. Because of Saturn's slow progress across the sky, and its sudden reversal of direction, Saturday, the last day of the week was named after it.

Although observed and recorded as a wandering star, the rings of Saturn were not seen until Galileo used his telescope to view the planet in the year 1610. At that time it was the furthest known object from the Earth. Galileo thought he saw "bulges" on its sides, but his telescope was not adequate to see Saturn's ring system. Galileo did discover several of Saturn's satellites, which he called planets. Since that time many great astronomers learned much about Saturn. Its exact size, motions, atmosphere, ring system, satellites, and so forth are now known. Three fly-by unmanned spacecraft were sent to Saturn in 1979, 1980, and 1981.

Early astronomers had the misconception that Saturn was a small planet. Their concept of the importance of the Earth was destroyed when they learned that about 765 Earths could be placed inside the confines of Saturn. Saturn is the next largest planet to Jupiter. It is considered a giant outer planet with an extensive and complex atmosphere with a density less than water. Below a thin cloud layer there is an extensive mixture of mostly hydrogen and helium with traces of methane, ethane, acetylene, along with traces of sulfur and carbon. Lightning has been observed there. The extreme pressure produces high internal temperatures and the gases become liquefied. Finally, at the center, they become solid metals. The surface atmosphere moves in jets of wind, both east and west at over 1,000 miles per hour. Small eddies and spots have been observed. One known as the Great White Spot comes into view every 29 years. It is thought to be crystallized ammonia.

Saturn's most unusual characteristic is its system of rings. Christian Huygens (1629–1695), the famous astronomer who invented the pendulum clock and proposed the wave theory for light, was the first to identify Saturn's ring system in either 1655, 1656, or 1657. Huygens also discovered Titan, Saturn's largest satellite. Only Ganymede, a satellite of Jupiter, is larger than Titan. In 1675 Giovanni Domenique Cassini (1625–1712) was able to make out two distinct rings around Saturn. In 1857 James Clerk Maxwell (1831–1879) demonstrated mathematically that the rings were made up of unconnected smaller particles, all revolving within the equatorial plane of Saturn. Recent analysis indicates that there are many bands of rings that range from 0.5 to 1.2 miles in thickness. They are made up of ice-covered silicate rocks. There is speculation that the origin of the rings were either debris from a comet that entered Saturn's gravitational field and broke up, or that one of Saturn's natural satellites was pulled in too close to the mother planet and disintegrated into billions of tiny rocks. Titan, the largest of Saturn's satellites, is the only moon in the solar system with a significant atmosphere of its own. On October 15, 1997 the United States sent Cassini, an unmanned spacecraft, on a 2-billion-mile trip to explore Saturn and Titan.

Uranus

Size: Uranus's equatorial diameter is 31,760 miles.

Mass: Its mass is about 14.55 times Earth's mass.

Motion: One revolution around the sun = 84.012 Earth years with a slightly eccentric orbit. Its period of rotation on its axis is approximately 17 hours and 55 minutes. Its axis is inclined at almost 98°.

Distances: Its average distance from the sun is 1,783.1 million miles (1.78 billion miles).

Temperature: Its temperature by infrared measurements = −355°F. It has no high internal temperature as do other planets.

Satellites: Uranus has five major natural satellites discovered by telescopes, plus about 10 smaller satellites that were discovered by the unmanned spacecraft Voyager 2. Others may yet be discovered.

Because Uranus is only visible on very clear, dry nights as a very faint point of light, it was not easily observed by ancient astronomers. Its brightness is only a magnitude of 5.5, which makes it almost impossible to see from Earth. Uranus (or Ouranos), the son of Gaia, was the major Greek god of all the sky and heavens. Because the Romans had no equivalent god in their mythology, Uranus was not given a Latin name.

Uranus was discovered in 1781 by Sir William Herschel (1738–1822), who also discovered the true motion of the sun through space, and the nature of the Milky Way. He described what he saw as a blue-green disk. There was a minor controversy over naming the new planet. Herschel wanted to call it "Georgium Sidus" after King George III. The French wanted to call it "Herschel," and finally Johann Bode proposed the name "Uranus" because of its place in Greek mythology (the sky god). About 100 years later this became the accepted name.

It is the seventh planet from the sun and is considered one of the four gas giants, along with Jupiter, Saturn, and Neptune. There are few recent misconceptions about Uranus, but there is one oddity: Its axis is inclined almost 98° from its orbital plane. This is unusual because it means that its poles are almost lined up with and pointing to its orbit around the sun. This is unique for a planet in our solar system because it causes Uranus's rotation to be retrograde. Even so, its magnetic field is inclined only 60° to its axis, which makes the planet's magnetic forces unequal at different locations.

Most of what we know about Uranus is from data sent back to Earth by the unmanned spacecraft, Voyager 2, which came within 50,600 miles of the planet's upper clouds. Its atmosphere is mainly hydrogen, with about 15% helium, which is a greater proportion of helium than the other gas giants have. Other gases and elements make up only 1% of the planet by weight. There seems to be a 6,000-foot-deep hot-water ocean under the clouds of Uranus. Its bluish-greenish color of reflected sunlight from its surface is the result of the absorption of red and longer wavelengths of light by its atmosphere.

It was not until 1977 that a ring system was discovered around Uranus. This system is made up of over 100 separate concentric rings mostly composed of dust particles. Some of the subrings include small rocks up to 8 inches in diameter. Uranus's moons are synchronous—that is, they keep the same face toward the planet. In the future they are expected to provide evidence of how the planet was formed.

Neptune

Size: Neptune's equatorial diameter is 30,775 miles.

Mass: Its mass is 17.2 times that of the Earth's. Its density is 1.7 g/cm^3 (water = 1.0).

Motion: Its period of revolution around the sun = 169 years. Its period of rotation on its axis at its south pole is just over 16 hours. Its period of rotation at the equator is about 18 hours. Neptune's axis is inclined to its elliptical plane at 28°48′, which is close to Earth's inclination of 23°30′.

Distances: Its mean distance from the sun = 2,794.0 million miles or 30.06 AU.

Temperature: The surface temperature is about –360°F, with increasing temperatures toward the center of the planet.

Satellites: Neptune has eight natural satellites, six of which were discovered by the unmanned spacecraft Voyager 2. Other smaller satellites are possible.

The brightness of Neptune is only a magnitude of 7.8, which is much too dim to be seen by the unaided eye, or even with a small telescope. Therefore there are no ancient misconceptions about Neptune as there are no records of ancients seeing this planet, or of Greek/Roman mythologies relating to it. There is unconfirmed speculation that Galileo may have seen it but did not recognize it as a planet. The story of its discovery is an example of independent astronomers calculating that there must be an eighth planet, located beyond Uranus. It seems that both John Couch Adams (1819–1892) and Urbain Jean Joseph Leverrier (1811–1877) saw slight perturbations of Uranus' orbit, meaning that its orbit showed irregularities because of gravitational influences by another unknown planet. Unknown to each other, they both reported their predicted discovery of a new planet. In 1846 Johann Gottfried Galle, along with Heinrich Louis d'Arrest, located the new planet almost exactly where it was predicted to be. This caused an international argument between France and England, concerning not only who discovered the planet, but also who should name it. The discoverers wanted to name it after themselves, but credit was finally given to both Adams and Leverrier because their calculations made the discovery possible. It was finally named Neptune after the Roman god of water and the seas because of its bluish color.

Most of what we know about Neptune was learned from data sent back by the unmanned spacecraft Voyager 2, which made its close fly-by (3,100 miles) on August 20, 1977. There are several cloud levels on the surface, some changing very rapidly. There is a Great Dark Spot (similar to Jupiter's Great Red Spot) named Scooter, located in a storm system above the surface. This spot is about the size of Earth. The wind speeds on Neptune are over 700 miles per hour. Its atmosphere is mostly hydrogen and helium, with a small percentage of methane, ethane, ammonia, and traces of carbon and oxygen. Neptune's orbit is the closest to being a circle of all the planets. Its orbit has an eccentricity (off center) of only 0.0086. Its internal temperature indicates that Neptune has an internal source of heat similar to Jupiter and Saturn that Uranus lacks. Some of its larger satellites have very unusual orbits as they have greatly accentuated elliptical paths. It largest satellite, Triton, is about the size of our moon. Neptune has two bright and several narrow rings surrounding it. They are not as extensive as the ring system of Saturn. For many years it was thought that Neptune was the final planet in the solar system, but this was not to be.

Pluto

Size: Pluto's diameter is 1,420 miles. (Earth's moon's diameter = 2,160 miles.)

Mass: Its mass is 0.002 of the Earth's mass. Its density is 2.1 g/cm^3.

Motion: Its period of revolution around the sun is not consistent but is about 248 Earth years. Its period of rotation on its axis (1 day) is about 6.5 Earth days. Pluto is inclined to its orbit at about 122°.

Distances: At its closest to the sun it is 2.75 billion miles away; at its farthest point from the sun it is 4.60 billion miles away. It has a large eccentricity of 0.249.

Temperature: Unknown, but assumed to be very cold.

Satellites: Pluto has one natural satellite, Charon.

Like Neptune, Pluto's existence was predicted by mathematical calculations based on the perturbations (irregularities) of the orbits of other planets before it was discovered. Oddly, the calculations were incorrect, but it was located just the same. Percival Lowell (1855–1916) searched for it from his private observatory in Flagstaff, Arizona. Pluto, as the ninth, and most distant planet from the sun, was discovered by Clyde W. Tombaugh (1906–) on February 18, 1930. Pluto and its satellite, Charon, are thought to be leftovers from the early formation of the solar system. In some ways, Pluto and Charon do not fit the pattern of the other planets. Most of what we now know about Pluto has been derived from three sources: (a) telescopic and spectroscopic observations; (b) theories about what we know of the characteristics of planets and the solar system; and (c) computer simulations. It is the smallest of all the planets and has the most eccentric orbit of all. Pluto's lopsided orbit even extends past Neptune's on its closest approach to the sun. There is no danger of the two planets colliding as their elliptical orbits are inclined at much different angles. Pluto is most likely composed of a core silicate rock with a diameter of about 500 or 600 miles. Outside the core is a layer of frozen ice, and on the surface there is a layer of methane frost. There is a definite color difference in the light reflected from the equator (more reddish) as opposed to that reflected from the poles (more bluish). Pluto's atmosphere is not completely known because it is so thin. It becomes thicker when closest to the sun, and thinner as it recedes from the sun. The atmosphere seems to be composed mainly of methane, nitrogen, ammonia, and carbon monoxide. Charon, Pluto's single satellite, is 721 miles in diameter. When viewed through a telescope, Charon appears as a dirty grayish fuzzy disk. It was discovered in 1978 by James W. Christy, an American astronomer. Charon's orbit around Pluto is about an average of 11,000 miles in diameter. It also has a small trace of an atmosphere.

Because of some perturbations causing irregularities in Neptune's orbit, it was speculated that another planet, possibly three to five times the size of the Earth, was roaming in a very distant orbit. The period of revolution in a very eccentric orbit around the sun was estimated to be about 1,000 years. After the data about Neptune was returned to Earth from Voyager 2, the existence of such a planet or planetoid was discounted. More recent data indicate that Pluto may not be a "real" planet, but rather a large asteroid within the Kuiper belt of debris left over after the formation of the solar system. The debate about the status of Pluto continues among astronomers.

1996TL66

In June of 1997 the Associated Press published an article indicating the discovery of an icy miniplanet found at the farthest reaches of the solar system. This planetoid (little planet) is 300 miles in diameter and is the most recent discovery in our solar system since Pluto's satellite, Charon, was exposed in 1978. Its eccentric orbit extends three times the distance from the sun as compared to Pluto's orbit. Jane Luu, an astronomer at the Harvard-Smithsonian Center for Astrophysics in Cambridge, Massachusetts discovered the new object, named 1996TL66, with the cooperation of other astronomers in Hawaii, Arizona, and at Harvard. This Texas-size chunk of matter is located in the Kuiper belt located just outside the orbits of Neptune and Pluto. This leads to the question: Is this a mini planet, an asteroid, a large meteor, or possibly a large comet? Its orbit around the sun is highly elliptical with a period of about 800 years, but it does not vaporize its surface leaving a tail as it passes the sun as do standard comets. It is not considered a major planet, but it is considered one of many millions of smaller objects that exist in the outer reaches of our solar system. It is very dark in color and is believed to be composed of methane and other hydrocarbons, carbon dioxide, and water, all of which are frozen.

The Sun and Other Stars

Sun

Let us begin our discussion with a few current, known facts about the sun.

Size: The sun's diameter is 864,000 miles; about 109 times the diameter of Earth.

Mass: Its mass is 1.99×10^{30} kg; about 333,400 times the mass of the Earth. The sun contains 99.86% of all the mass in the solar system. Its average density is 1.4 g/cm^3.

Motion: The sun is inclined to the ecliptic 7°. The period of surface rotation at the equator is 25 days, which is faster than at the poles, which is 36 days. The rate of rotation for the inner layers of the sun is not known. The sun, along with the entire solar system, is moving at about 12 miles per second through space (celestial sphere) in the direction of the star Vega. No need to worry! It will take billions of years to reach Vega, which by that time, may itself have moved.

Distances: Its mean distance from Earth is 93,000,000 miles (1 AU); its closest distance to Earth = 91,400,000 miles (in January); its greatest distance from Earth is 94,500,000 miles (in June).

Gravity: The sun's surface gravity is about 38 times Earth's gravity. (A 100-lb person on Earth would weight 3,800 pounds on the Sun's surface.) Because of the great internal pressure from its size, 1 cubic inch of matter from the sun's core would weigh about 100,000 tons on Earth.

Composition: Hydrogen and helium account for 95% of all the elements comprising the sun. More than 60 elements found on Earth have been identified on the sun by spectroscopic analysis. The sun is composed of different layers, each with its own characteristics and temperatures:

Core: The diameter of the core of the sun is 249,000 miles; its temperature is 27,000,000°F; its pressure is about 7 million pounds per square inch. The core is where thermal nuclear fusion takes place converting hydrogen into helium and producing the sun's light, heat, and energy.

Photosphere: The photosphere's diameter is 500 miles; its temperature is about 9,800°F. The word *photosphere* means the "sphere of light." This layer is mottled in appearance, which is caused by granulation. It is the region where light is sent out from the surface.

Chromosphere: The chromosphere's diameter ranges from 2,000 to 6,000 miles; its temperatures range from a relatively cool 7,200°F to 90,000°F. The layer is known as the color sphere as it provides the reddish glow of the sun.

Corona: The corona is a thin layer of ionized gas that is only viewed during an eclipse of the sun. It is about 46,000 miles above the surface, and has a much higher temperature than does the chromosphere. Temperatures reach 3,600,000°F. (Note: all figures are approximate.)

Sunspots: The sun has a high magnetic field that is wound around the sun rather than aligned north to south, as is the Earth's magnetic field. Sunspots are giant loops of magnetic flux, which range from about 3,000 to 62,000 miles in diameter. This magnetic loop activity causes temporary dark areas on the surface of the sun. They become solar flares that emit radio waves that leap out from the sun. These charged particles affect electronic communications on the Earth. They may last from a few days to several months. The sun has a 22-year cycle, whereas the sunspots exhibit two 11-year cycles during this period. Periods for sunspot cycles vary over history.

The sun was the first object in the sky viewed by human beings. Its powerful light, its regular disappearance and reappearance must have confused and fascinated people from the beginning of time. No doubt this is why so many misconceptions, beliefs, myths, and religions developed about the sun through the ages. Ancient myths about the sun resulted from misconceptions about the solar system and other heavenly bodies. Such beliefs and misconceptions were abundant.

The power of the sun impressed the Egyptians. Re (also known as Ra) was their major sun god. The Egyptian creation myth relates that Re emerged from primordial water by self-fertilization. Each dawn the sun was greeted as the birth of a newborn child from the womb of Nut, the sky goddess. At noon the sun was a flying bird, and at dusk it was an old man who retired underground in the western sky. About 2000 B.C. the myth was that the sun was attacked at sunset, a battle continued all night, the sun won and thus would rise again in the east. Part of this myth is related to the Egyptian scarab beetle, which is a species of dung beetle. The Egyptians believed that the dung beetle used its legs to roll the fiery ball of the sun on its daily trip across the sky, as the beetle rolls its dung ball.

The Greeks attributed Apollo, a child of Zeus and Leto, the powers of the sun god. He assumed a great many duties beyond that of sun god, including presiding over music, poetry, medicine, beauty, law, archery, and prophecy. In later Roman mythology he was still called Apollo, but also Phoebus.

In Bantu mythology, Wele (chief god) made the heavens, including the two brothers: the sun and moon. Because they did not know what the moon and sun were, or how they moved, they created stories to explain their observations, as did most ancient civilizations. One story relates that the moon kicked brother sun out of the sky. To get back at the moon, sun knocked brother moon to the earth, where he became covered with mud, dimming his brilliance in the night sky. This dispute was settled when Wele said that the brothers should never be in the sky at the same time.

The Arctic Eskimo's misconceptions about the moon and sun are expressed in a more elaborate story. It deals with the sun, the pretty sister with a bright face, and her brother, the moon, who was promiscuous and had a dirty face. Moon covered his face with ashes so that he could fool his sister and make love to her all night long. Being pure brightness, when the sun recognized the moon as her brother, she was upset because incest was forbidden. She mixed her blood, which flowed from a breast the moon had cut off, with her urine and told her brother that he could drink it if he desired her so much. Sun then took a torch of fire and jumped into the sky. Brother moon waited and then followed with a weaker torch, but did not catch sun. This affair continues nightly. The sun runs from east to west across the sky, with brother moon always chasing her.

The Aztecs made sacrifices to the sun and seasons, hoping to improve their well being. The equinox celebrations stood for "flaying of men"; to symbolize this the priest dressed in the skin of a flayed sacrificial victim. This skin was likened to the coat of a seed, which must be shed for the plant to grow. From the dead skin new life emerges. The ancient Mexican cultures made many other types of human sacrifices. Hearts of sacrificed victims were held up to the east as food offered to the sun. They ritualized life and death as part of nature.

The Maui's of Samoa had the misconception that a mythical being put a rope noose around the neck of the sun to catch it. These ropes were the rays of sun seen at dusk and dawn; they were used to lower the sun at sunset, and draw it back up again in the morning.

The Cahto Indians of California have a sun myth that involves a misconception about the powers of coyotes over the sun and seasons. As the coyote sleeps, he moves his head in the four directions. Asleep with his head pointed east, he gets warm and brings the sun back for the people.

There are many myths relating the sun to the seasons. Many of them were based on the fear that a setting sun would not return, so people had to act to ensure that it would. Ancient people believed the sun and moon caused the seasons.

The Maori of New Zealand were aware of the solstices and the seasons. During their summer, the sun is at Rangi's head, and progresses to his toes until the winter solstice (June 21), when he spends time with his winter wife, the Winter Maid. During the summer solstice in New Zealand (December 21) the sun is spending time with his Summer Maid, who raises the crops.

Old English misconceptions about the sun led to a variety of superstitions. For example, the day of the week and time of day for births is important in English superstitions. It was agreed that a child born on Sunday morning (the sun day) would be lucky, gifted, and free from evil spirits and death. Thus the saying, "The child born on the Sabbath Day is blithe and bonny, good and gay." A person who pointed at the sun, moon, or stars would have bad luck. In ancient custom, one would turn toward the sun (called *deiseil*) for good luck when doing something important. Movements and dances always turned sunwise (clockwise). The opposite direction, countersunwise, brought bad luck. Millstones moved sunwise, as did all clocks. When cooking, one stirred in a sunwise direction. Fishing vessels made three *deiseil* turns when leaving port to assure good catches. One always walked *deiseil* around a respected person. Large bonfires were set on the first day of summer to enhance the sun and ensure a good crop. This practice was also carried out at the autumn equinox and became Halloween.

Many ancient stories tell of fires starting as sun passed through crystal or glass. In the first century A.D. an Indian mystic used a ring of glass to concentrate the sun's rays to start sacred fires. As recently as 1985, the King of Thailand lit the Queen's funeral pyre using a crystal to focus the sun's rays. Burning glasses were used to harness the power of the sun for centuries. Such a glass was found in Crete dating back to 1,500 B.C.

Over the centuries people used the sun, moon, and stars to keep time. The regular motions of these bodies were evident and thus used to determine the appropriate days for planting and conducting religious ceremonies.

A little-known fact is that in about 280 B.C. the Greek astronomer Aristarchus of Samos (c.320–250 B.C.) conceived of a sun-centered universe hundreds of years before Copernicus. Aristarchus not only placed the sun at the center of the universe, he also explained the daily rotation of the Earth on its axis and its yearly revolution around the sun. He stated that all the planets revolved around the sun and that the apparent movement of the "fixed" stars was caused by the Earth's movements. He also considered the universe to be much larger than anyone else thought possible.

Regardless, misconceptions about an Earth-centered universe continued. For thousands of years most astronomers considered the Earth to be fixed at the center of the universe. The most famous proponent of this geocentric model is Ptolemy of Alexandria (c.90–170), whose conceptions were based on Aristotle's' misconceptions. His model confused things for 1,500 years (see Figure 5.1). It was not until 1514 when Copernicus, using Plato's ideas, developed his solar-centered model, finally published in 1543, that Ptolemy's misconception was corrected (see Figures 5.2 and 5.3). Our knowledge of the sun was limited until the early 1600s when Galileo observed sunspots and determined that the sun rotated on its axis based on the movement of the sunspots. In the late 1600s, when in triangulation with Mars, the distance from the sun to the Earth was established. In 1814 Joseph von Fraunhofer (1787–1826) discovered the absorption bands in the sun's spectrum. He identified over 600 spectral lines using the new diffraction grating and an instrument called the spectroscope. Gustav Robert Kirchhoff (1824–1887), who also made several contributions to the field of electricity, determined its chemical composition and some physics of the sun by analyzing the sun's spectrum. In the early 1900s George Ellery Hale (1868–1938) determined the source of the sun's energy as the thermonuclear fusion of hydrogen. He also determined that sunspots were caused by the sun's magnetic field.

Beliefs and misconceptions about solar and lunar eclipses have always existed. A lunar eclipse occurs when the Earth casts a shadow through which the moon passes. For this to happen the moon and Earth must be aligned with the sun. Solar eclipses are less frequent. They occur when the moon passes across the face of the sun when the Earth, moon, and sun are aligned. A total solar eclipse can only occur on a new moon, when the moon is directly in the path of the sun and blocks its image. Even though the sun is about 400 times the diameter of the moon, a total solar eclipse is possible because the moon is about 400 times smaller than the sun and is much closer to the Earth than the sun. A total solar eclipse lasts about 7 1/2 minutes.

To the ancients, a solar eclipse meant trouble. It was seldom taken as a good omen. The term chosen to describe this phenomenon indicates the unease the ancients felt at its occurrence. The word *eclipse* is derived from the Greek word *ekleipsis,* meaning "abandonment." During a total solar eclipse the daylight becomes ruddy and takes on an eerie glow. The Chinese understood what caused an eclipse as early as 200 B.C. After astrologers were able to estimate the periods of revolution for the moon and sun around the Earth (to them the Earth was still the center of the world), they could approximate the dates of reoccurrence.

In China, an account of a solar eclipse was recounted in the Confucian classic, *Shih Ching,* in the sixth century B.C. The Chinese word for eclipse is *shih,* which means "to eat." It was thought that a dragon dined during an eclipse. Thus, an eclipse was ugly, dirty, and abnormal. The Chinese also believed that during a solar eclipse the sun was eaten by the heavenly dog. The inhabitants of the Baltic countries were convinced that giant serpents, dragons, or witches were trying to devour the moon or sun during eclipses. Ancient people of various cultures used the eating metaphor to explain such phenomena. These misconceptions, beliefs, and myths related to lunar and solar eclipses are only a sampling of the hundreds that exist.

The Babylonians were careful observers and recorders of both lunar and solar eclipses. They used their records to predict that an eclipse would occur every 18 years. Assyrian astrologers determined that lunar eclipses would occur about every 41 or 47 months. The differences result from calculation methods. Old astrological records indicate an awareness that the less frequent solar eclipse occurs about 6 months before or after a lunar eclipse.

The ancient Greek philosopher/astronomer Anaximander (611–645 B.C.) proposed that a flat earth floated in a cylinder, suspended in endless space. The sun, moon, and stars were hollow wheels of fire in the cylinder above the earth. They emitted light through a small opening. During an eclipse this small opening closed. Other ancient Greeks believed that the sun hid behind mountains and did not go below the flat earth at night, but that the stars did. Others believed the sun became invisible at night because it went so far away from the earth and then would reappear in the morning when it came close again. They had no rational explanation for eclipses, except that something went wrong with the system, and therefore something would go wrong on Earth. Hipparchus was the first to use geometry and a solar eclipse to calculate the parallax of the moon. Using mathematics, he determined the moon's distance from the Earth. His estimate ranged from 60 to 75 radii of the Earth, based on the time of the year.

On one of his voyages to the New World, Columbus used tables listing predictions of eclipses to his advantage with the natives. Knowing that a solar eclipse would occur on February 29, 1504, Columbus manipulated the inhabitants of one of the islands. Other stories exist about the use of solar eclipses to influence the uninformed.

Stars

There is some confusion as to the distinctions between *astronomy* and *cosmology*. Generally speaking, astronomy is the science of the celestial bodies of the universe, including the solar system, as well as stars, galaxies, and interstellar matter. Modern cosmology is theoretical. It is based on universal physical laws with mathematical underpinnings. Cosmologists study the origin, structure, composition, and evolution of the entire universe. Cosmology might be described as the theoretical explanations of the creation and nature of the universe, its evolution and its future. Astronomy began when people first gazed at the night sky. Cosmology began in the 1600s with Copernicus's heliocentric theory and Galileo's first glimpse through a telescope. We will next consider scientific developments and misconceptions about the stars, galaxies, and universe after examining some current facts.

Star Sizes: The largest stars have a diameter over 400 times that of our sun. The sun's diameter is 864,000 miles, which is about 110 times the Earth's. Small white dwarfs (see Table 5.2) have a diameter of only 0.01 compared to the sun.

Star Masses: The largest stars have the lowest densities. They consist mostly of gases. Superstars have a mass 1,000 times that of the sun, whereas the tiny white dwarfs are small but extremely dense.

Star Classifications: Stars can be classified in several ways: brightness, distance, spectroscopic characteristics, or color, which is related to temperature and the intensity of their hydrogen spectral lines. See Table 5.2 for some examples.

Table 5.2
EXAMPLES OF STAR CLASSIFICATIONS

Class	Color	Temperature	Name
O	Blue	45,000–75,000	Epsilon
B	Blue	21,000–45,000	Rigel
A	Blue-White	14,000–21,000	Sirius
F	White	11,000–14,000	Polaris
G	Yellow	9,000–11,000	Sun
K	Orange	6,000–9,000	Arcturus
M	Red	5,000–6,000	Betelgeuse

Star Magnitudes: Stars may be arranged by their brightness as viewed from Earth with the unaided eye. The brightest stars are over 1,000,000 times brighter than the sun, whereas the white dwarfs are about 1,000 times less bright. To earthly observers, Sirius is the brightest star (excluding the sun). It is located in the constellation Canis Major with an apparent magnitude of −1.47. One of the dimmest stars is Betelgeuse, the red giant, located in the constellation of Orion. It has an apparent magnitude of +0.85.

Star Distances: A star's distance from Earth is measured in light-years, the distance light travels in 1 year at a speed of 186,000 miles per second, or 5.879×10^{12} miles per year. The astronomical unit (AU)—the distance from the Earth to the sun (or 93,000,000 miles)—is used to measure large distances. The parsec (parallax/second) is also used to measure great distances in space. One parsec equals 206,265 AUs, or 3.258 light-years. Table 5.3 provides examples of stellar distances, star types, and magnitudes (brightness). The distance of stars relatively close to the Earth can be measured by parallax by taking measurements every 6 months as the Earth revolves about the sun. A more exact method is to use the Doppler effect, which is based on the color shift (frequency of light) related to the star's velocity. The light frequency for a star receding from the Earth is redder (longer wave length) than a star approaching the Earth, which emits a blue light (shorter light wave length). The same Doppler shift occurs with sound waves, when, for example, a train whistle becomes higher in pitch (frequency/shorter sound waves) as it approaches you, but lower in pitch (frequency) as it recedes from you.

Table 5.3
STAR QUALITIES

Star	Light-Years	Type	Apparent Magnitude
Sun	(1 AU)	G2	−26.73
Proxima Centauri	4.3	M5	+10.75
Alpha Centauri	4.3	G2	0.0
Sirius A	8.7	A1	−1.47
61 Cyg. A	11.2	K5	+5.2
Procyon A	11.3	F5	+0.3

Types of Stars:

A *white dwarf* is in its last stages of evolution, when it has exhausted its supply of nuclear fuel. It collapses under its own gravity and becomes very dense, which produces the heat that provides the bright glow. An example is Sirius B, the companion star to Sirius A.

A *black dwarf* is the final stage of a star after it has used up the energy of the white-dwarf stage. It emits no light.

Black holes are theoretical. They are thought to be vortex areas in space where massive stars have collapsed, creating such great gravity that no light can escape into space.

White holes are also theoretical—one has never been seen, but they are thought to be where new star matter is created spontaneously.

Hot subdwarfs are hot blue stars with high densities.

Blue supergiants are the bluest, hottest, and brightest stars. They contain very large masses of matter, but at low densities. Rigel, part of the constellation Orion, is an example.

Red/orange giants have low densities and are much larger and brighter than the sun. Arcturus provides an example.

Red supergiants are the largest in size, but have low densities. The best known red supergiant is Betelgeuse, found in the constellation Orion.

Neutron stars are the remnants of stars that have imploded with enough energy to force particles together to form neutrons.

Pulsars emit short bursts of radio waves. They are thought to be neutron stars that are rotating.

Binary star systems (composed of two stars) shut out each other's light as one eclipses the other. An example is Sirius with its companion star in the constellation Canis Major.

Variable stars change their magnitude (brightness) from time to time (see *binary stars*).

Cepheid variables are a pair of pulsating stars of similar brightness that regularly and periodically eclipse each other.

Supernova is a great explosion of a large star that collapses because of its gravitational force, sending great bursts of light into space. One was detected in the Crab Nebula in the year 1054, another in 1987 in the Large Magellanic Cloud.

Constellations: The conspicuous patterns or groupings of several stars are given different names by various civilizations. Most constellations were named by the Arabs and Greeks and are identified by their abbreviations. In Greek, "zodiac" means "ring of animals." Although constellations are usually given Latin names, the individual stars that make up the constellations are given Greek names. Currently, there are about 88 recognized constellations. Table 5.4 lists the 12 constellations of the zodiac. Particular planets are located in specific "houses" and astrologers believe they are associated with human traits and events.

Table 5.4
CONSTELLATIONS OF THE ZODIAC

Name	Dates	Symbol
Aries	March 21–April 19	Ram
Taurus	April 20–May 20	Bull
Gemini	May 21–June 20	Twins
Cancer	June 21–July 22	Crab
Leo	July 23–August 22	Lion
Virgo	August 23–Sept. 22	Virgin
Libra	Sept. 23–Oct. 22	Scales
Scorpius	Oct. 23–Nov. 21	Scorpion
Sagittarius	Nov. 22–Dec. 21	Archer
Capricornus	Dec. 22–Jan 19	Goat
Aquarius	Jan. 20–Feb. 18	Water Bearer
Pisces	Feb. 19–March 20	Fish

The early Babylonians, Chinese, Indians, Mayans, and later the Greeks and Egyptians, all kept records of the brightest stars and maintained maps of the constellations. They believed that the constellations represented animals or people who could affect events on Earth. Thus was born the idea that the heavenly bodies have some influence on the fate of people. A Greek poem by Aratos (or Aratus), written sometime after 600 B.C., popularized the names of the constellations so they could be remembered and used by farmers and others. This poem may be one reason why the Greek names for the constellations became popular and are used today rather than the Arabic, Chinese, or Egyptian names.

Galaxies: Galaxies are huge groupings of million or billions of stars held in one of several shapes by their mutual gravity. There are elliptical, irregular, and spiral types of galaxies. The Milky Way galaxy has a typical spiral shape, with our solar system located about a third of the way toward the center on one of its spiral arms (see Figure 5.7). The Milky Way is a flattened disk that contains about 100 billion stars. Its diameter is about 300,000 light-years, and it is about 10,000 light- years thick. The Milky Way is also one of about two dozen other galaxies that form what is known as the Local Group, which is composed of over 700 billion stars. One estimate is that there are over a *billion billion* stars in the universe. In addition, the heavens are filled with extragalactic systems, nebulas, globular clusters, and other matter. The observation of other galaxies receding from each other in space led to the theory that space is expanding and infinite.

Figure 5.7 An artist's depiction of a spiral galaxy similar to our Milky Way.

Some ancient religions considered that the stars and planets represented human souls, others thought the stars were gods, and still others believed that great heroes were placed in the heavens and became stars. These early people believed that these objects controlled human events and activities.

Ptolemy of Alexandria listed 47 of the constellations in the northern hemisphere. Those in the southern hemisphere were not charted until the Middle Ages when explorers made maps of the southern skies. Tycho Brahe, Jakob Bartsch, Augustin Royer, Johannes Hevelius, and other astronomers organized the constellations into the 88 groups now recognized. In 1930 the International Astronomical Union established boundaries for the constellations so they can be used to indicate general direction in the sky for locating celestial objects.

Isidore of Seville (560–636) came up with many misconceptions about astronomy based on old Greek writings. He believed that the universe was of limited size, was only a few thousand years old, and was at the end of its life. He also believed that the universe would expire in his lifetime. Isidore considered the Earth to be wheel-shaped with five zones, only two of which were inhabitable. He thought the Earth was surrounded by water with spheres of stars overhead. He also believed that beyond the stars existed a heaven of the blessed. Isidore also held the misconception that what he called the "firmament" revolved around the Earth because he mistakenly contended that the Earth did not move. Bede (672–735) continued the misconceptions of Isidore, but Bede considered cause and effect when trying to explain his universe. Even so, Bede believed that the Earth was a static sphere surrounded by seven heavens: the firmament containing the planets and stars, the heaven of the angels, the heaven of the Trinity, Olympus, air, æther, and fire. It was a common misconception that the firmament was the outer boundary of the universe.

The Arab philosopher/astronomer Al-Kindi (c.800–875), as well as most of the philosophers of the Middle Ages, held the misconception that magic, the occult, and astrology were the physical causes of natural phenomena. He believed that rays given off by heavenly bodies affected the human mind as do words. As rays from the stars changed in the celestial spheres, so did events on Earth. Later, the alchemist Roger Bacon (1220–1292) and others before his time, including Al-Kindi, considered occult virtues (natural magic) that reside in the stars and other heavenly objects to be the cause of physical events. Roger Bacon's ideas were considered viable science in his day.

Most of the stars seemed fixed or unmoving in the night sky. There were visible exceptions however; an early misconception was that planets (the wanderers) were stars that were not fixed. As previously mentioned, Aristotle and others believed that the stars existed in the æther of crystal hemispheres above a flat Earth, or as a series of concentric spheres encircling a spherical Earth. Aristotle's misconception about the cosmos was that the universe was a series of large, finite crystal spheres (like the layers of an onion) with the Earth at its center. The fixed stars were in the outermost sphere, which he considered the "prime mover." This *primum movens* acted as the main source of all movement in the universe. Separate spheres held the sun, moon, planets, and different groupings of stars. Even the astronomers of the Renaissance, who accepted the heliocentric concept of Copernicus, still believed in celestial spheres and that planets moved in perfect circles.

The English astronomer Edmund Halley was the first to realize that the stars are not fixed to solid celestial spheres and that they actually do move. He realized that stars just appear to be "fixed" because they are much farther away from us than they seem to be and their motions are difficult to detect. In 1718 he observed the actual motions of the bright stars, Sirius, Procyon, and Arcturus. He estimated the distance of Sirius from the Earth as 2 light-years. His estimate was off because he assumed Sirius was the same brightness as the sun. Actually, it is much brighter than the sun and is 8.7 light-years distant. Halley was also the first to suggest that the "cloudy" matter in space, called *nebula,* was actually interstellar matter forming new stars.

Friedrich Wilhelm Bessel (1784–1846) perfected Halley's method of measuring star movement and distances. He used parallax (apparent differences in the positions of a star due to a change in the observer's position) to calculate the distance of the star 61 Cygni. It takes 11.2 years for light from this star to reach Earth. At that time, it was thought to be the closest star to Earth (with the exception of the sun). Later, a three-star system, Alpha Centauri, was discovered that was only 4.3 light-years away. One of the three, Proxima Centauri, which is about one twenty-thousandths as bright as our sun, is actually the closest to the Earth. Knowing the distance a star is from us is important in determining its real brightness, and the brightness determines a star's massiveness. The more massive a star, the greater is its gravity, and thus the more rapidly it will, through nuclear fusion, use up its hydrogen and collapse into itself. Thus in general the more luminous a star, the shorter its life. The sun, being an average star, has used up about half of its hydrogen. In about another 5 billion years it will become a red giant and expand to consume the entire solar system, and then collapse into a white dwarf before burning out.

The twentieth century has also seen misconceptions relating to the cosmos. A man named Charles Hoy Fort had some interest in science and wrote several books in which he duped many learned men about his beliefs. (His birthdate is unknown, but he published books in 1909, 1923, and posthumously after his death in 1932.) His books damned conventional experimental science as the lost souls of data. Fort's misconceptions included his belief that the Earth rotated once a year on its axis. To explain the motion of the stars around the Earth, he designed an opaque shell around the Earth that had holes in it to let the starlight shine through. The twinkling of the stars was caused by the quivering of this shell. Once in a while meteors penetrated the soft spots of this shell. Fort said he had many samples of a soft, gelatin-like substance from the shell. Nebulae were glowing light patches on the shell, whereas dark nebulae were opaque patches. He explained the famous horse-headed nebula as a refusal to mix with the other bright substances. Fort even proposed tours to the moon and stars. It was difficult for the press to determine if he was a great jokester, a nut, or a genius. He once said that his ideas were no more ridiculous than Darwin's, Newton's, or Einstein's theories.

Other Objects in the Universe

There are other objects in the universe that are visible from Earth besides the sun, moon, planets, and stars. Not all of them have regular, periodic motion. In this section we review the beliefs and misconceptions related to asteroids, meteors, comets, and aberrations in space.

Asteroids: Asteroids were never seen before the invention of the telescope. They are invisible to the unaided eye. The word *asteroid* is derived from the Greek word *asteroeides,* which means "starlike." That is how they appear when viewed through a telescope. They are somewhat similar to small planets. In fact, at one time they were called planetoids because they are small and revolve around the sun. Early astronomers observed and treated asteroids as minor planets the size of boulders. They measured their motions, mass, and other characteristics. As previously mentioned, the vast majority of asteroids are found in a planetary orbit called the *asteroid belt,* which is located between the orbits of Mars and Jupiter. The asteroid belt is about 2.1 to 3.3 AU from the sun. Their orbits range in size from about 2.3 to 2.8 times the Earth's orbit around the sun. Not all asteroids follow the same orbit. Their periods of revolution around the sun range from 3.6 to 4.6 years. More than 5,000 are known and possibly as many as another 5,000 exist but have not been detected. The total mass of all the asteroids is estimated to be about one one-thousandths that of the Earth's mass. Many of the larger asteroids have been given names. One group of 40, called *Apollo asteroids,* have orbits that intersect the Earth's orbit. In 1989 the asteroid Asclepius passed within about 400,000 miles of the Earth. In 1992 the asteroid No. 4179, Toutatis passed within 1.5 million miles of the Earth. Other small groups of asteroids intersect the orbits of Mars and Jupiter. One method of classifying asteroids is by their orbits in the area of the inner planets. For instance, *Aten* asteroids have orbits inside the Earth's orbit; *Amor* asteroids intersect Mars's orbit around the sun; *Apollo* asteroids cross the Earth's orbit; and *Arjuna* asteroids, which are very small—about 250 to 350 feet in diameter—have circular orbits around the sun.

There is another interesting group of asteroids, called *Trojans,* that are in the same orbit around the sun as the planet Jupiter. They have not been studied in great detail because at one time they were thought to be a cloud in Jupiter's orbit and were difficult to observe because they are dark. The Trojans are slow-moving asteroids that are about 200,000 in number; most are more than a kilometer in diameter. Oddly, one group travels in front of Jupiter in its orbit. These preceding Jupiter are named after the heroes that fought on the Greek side of the Trojan war, while the asteroids that follow Jupiter in its orbit are named after the heroes who fought on the Trojan side.

The origin of asteroids has been debated for centuries. One problem is that they are confused with meteors, small planets, small satellites of other planets, and even comets. This is because they are leftover material from the formation of the early solar system. They just did not coagulate by gravity or accumulate enough space material in early collisions to form larger bodies, as did the planets. One speculation is that they once were larger and broke into smaller pieces. The supporting evidence for this theory is their very irregular shapes. The three largest asteroids are Ceres, Pallas, and Vesta. It is difficult to obtain the exact measurements of the sizes of asteroids. Table 5.5 lists some facts about a few asteroids.

Table 5.5
ASTEROIDS

No.	Name	Year Discovered	AU to Sun	Diameter (mi.)
1	Ceres	1801	2.55	488
2	Pallas	1802	2.11	334
4	Vesta	1807	2.15	345
8	Flora	1847	1.86	94
10	Hygeia	1849	2.84	280
624	Hector	1907	4.99	91
944	Hildago	1920	2.00	9
2060	Chiron	1977	8.50	68

Another way to classify asteroids is by their chemical makeup, as inferred from the composition of asteroids. There are three types: (1) over 70% are *C-type asteroids,* which are made of dark carbonaceous rocks; (2) about 15% are classified as *S-types,* which are composed of grayish-stone material; and (3) over 10% are *M-types,* which are composed of various metals.

All planets and their satellites have been bombarded by asteroids since the formation of the solar system. This is how much of the land mass of planets was formed. The planets with atmosphere are somewhat protected, at least from the smaller asteroids, which burn up by friction as they pass through the atmosphere. This is why the moon has more visible craters than does the Earth. Even so, scientists have identified about 150 impact areas on the surface of the Earth. Most of them were caused millions of years ago by either comets, large meteors, or asteroids. A few craters seem more recent. One theory, which may become a misconception as study continues, is that past asteroid impacts caused the extinction of about half of all plant and animal species millions of years ago, as well as being the major cause of the ice ages. These chunks of rock are too small to interrupt the Earth's motion, but they could strike the surface with enough energy to explode into gas at very high temperatures and cloud the atmosphere with dust that would block the sun and cause great devastation to all living things. A related theory considers these effects to be responsible for the rapid evolution of new and different species of plants and animals (punctuated evolution). In the recent historical past, recorded asteroid impacts were much smaller than the scenario just described.

The Greeks, Egyptians, Chinese, and others reported fiery stones in the sky. In the Middle Ages and later, Europeans held many superstitions about asteroids, comets, and meteorites. One asteroid impact was reported, but the Paris Academy did not accept the report because they believed that it would encourage more superstition in the public. We are still confused as to whether these impacts are caused by asteroids or meteors. Their fiery entry into the Earth's atmosphere and the damage they both can cause makes the difference immaterial.

Meteors: Both scientists and the public have been confused by the distinction between *asteroids* and *meteoroids.* This may be because their compositions are similar as most meteors originate from the stony/metallic debris that results from the collisions of larger asteroids and possibly comets. Another theory is that meteors are comprised of matter left over from the formation of the solar system. A meteor is a small meteoroid, or a part of a meteoroid that has entered the Earth's atmosphere. Another confusing fact is that both asteroids and meteors enter the Earth's atmosphere, but meteors do so much more frequently. But there are several important distinctions between meteors and asteroids: Most asteroids, which are also considered small planets or planetoids, are larger than meteors, some being much larger. Another difference is that asteroids usually stay in the so-called asteroid belt between Mars and Jupiter, whereas meteors frequently enter the Earth's gravitational field and atmosphere. An arbitrary distinction is that meteors are not visible by reflected light when using even the best telescopes, whereas asteroids are.

When meteors enter the Earth's atmosphere, they become what we call *shooting* or *falling stars.* Similar terms have been used since prehistoric times. Some ancient cultures considered shooting stars good luck, but more often they were considered bad omens. Large meteors that produce spectacularly bright streaks of light are called *fireballs.* Once in a while a fireball splits into two or more parts: this is known as a *bolide,* after the Greek word *bolis,* which means "missile." A loud hissing noise can sometimes be heard as one of these large meteors passes through the atmosphere.

As the Earth travels in its orbit around the sun, its gravity pulls in all kinds of leftover matter from space, particularly when passing through the meteor belt. Most meteors caught up in the Earth's gravity are small particles, ranging from dust to chunks as large as trucks and houses. These particles travel about 35 miles per second as they enter the atmosphere. At this speed, enough friction is caused to heat the air surrounding the particles to over 3,500°F, thereby producing the shooting-star effects. The smaller meteors are completely vaporized, possibly leaving ash or dust behind, which become the nuclei that provide a base for water vapor to condense and form rain. Over 90% of all meteors entering the atmosphere never reach the Earth. When a larger meteor does not completely vaporize, a portion of it lands on the surface of the Earth. These surviving meteors are called *meteorites.* It has been estimated that over 50 tons of meteorites reach the Earth's surface every day. Most meteorites land in the oceans.

There are several ways to classify meteorites: *siderites* are iron meteorites composed of 90% iron and 10% nickel; *aerolites,* the most common type, are stony meteorites that are made up of about 90% silicates and 10% iron–nickel; *siderolites* are stony iron meteorites that are composed of 50% silicates and 50% iron–nickel; and *tektites* are small, dark, glasslike rocks found mainly in Australia and Thailand.

Recent evidence indicates that some meteorites originated on Mars. They were kicked into space by massive impacts on the surface of Mars, and possibly the moon. So far, eight Martian meteorites have been found. Their composition is different than other meteorites.

One can only wonder how prehistoric people interpreted the shooting bolts of light in the night sky. Some Stone Age axes seem to have been formed from iron meteorite material. This was many years before people learned to smelt iron from its ore. Throughout recorded history civilization considered meteor showers a cause for fear.

Homer's poem, the *Iliad,* relates a story in which lumps of pure iron were given as prizes at tournaments and contests. Because iron was not commonly smelted by the ancient Greeks, these gifts are thought to have been from meteorites. About 200 B.C. a large stone that fell from the sky was erected in the temple of Artemis. The Greek astronomer Hipparchus in the second century A.D. reported viewing a large falling star, which he interpreted as a sign from heaven. The temple of Kaaba in Mecca contained a black stone that was believed to be an iron meteorite. The Roman Pliny the Elder wrote in his famous compendium about "thunder stones," which he likened to axes from the sky. In later centuries the Chinese and Mongols referred to "lightning stones," which they used to make knives, axes, and other tools. Iron meteorites were highly valued. The high nickel content of iron meteorites made the iron hard and well suited for forming weapons and tools.

Old England made a connection between falling stars and the human soul. A falling star foretold a birth, the passing of a soul to or from heaven, or a soul's release from purgatory.

One does not need to go back to antiquity to find odd beliefs and misconceptions about meteors. In 1807, while president of the United States, Thomas Jefferson (1743–1826) was informed by two Yale astronomy professors of the discovery of a large meteorite. Despite Jefferson's reputation for being a scientist, he is said to have responded, "I would rather believe that two Yankee professors would lie than that stones fall from the sky." It was not until 1950 that scientists finally accepted the concept that an object large enough to form the giant meteorite crater in Arizona could fall from the heavens. Up to that time, the misconception was that the crater was formed by a huge gas bubble that came from inside the Earth. Another rather modern misconception relates to a giant crater found in South America, which still had the meteorite at the bottom of the hole. Geologists claimed that the hole could not have been caused by an object from the sky, but rather that the hole was dug by local Indians who buried their sacred iron objects in it.

In 1996 an international team of scientists studying deep-sea sediment off the coast of South Carolina discovered evidence of the impact of a very large meteorite. The ancient meteor landed near the Caribbean about 65 million years ago. After examining drilled core samples, scientists located a 3–8-inch layer of debris that was composed of material ejected into the ancient atmosphere when the meteorite impacted. This extensive cloud of dust and gas is thought to be the cause of global changes that blocked out sunlight, leading to the mass extinction of microorganisms, plant life, and the dinosaurs at about 65 million years ago. It might be noted that seeds, small animals, and insects could survive such tough times better than could large animals. This evidence corresponds with the discovery of the rare element iridium, which was found in sedimentary rock and may have been formed by the compaction of meteorite debris. The level of iridium found is 25% higher than the level that appears in the Earth's crust. Geologists estimate the age of this concentrated iridium at about 65 million years.

Few humans realize the importance of ancient asteroids and meteorites to our civilizations. They are the source of almost all of our minerals and heavy metals, including iron, cobalt, nickel, silver, and even gold. When the Earth formed, all of the heavy elements became concentrated in the core, leaving mostly lighter elements, silicates, rocks, and some minerals on the surface. The surface of the Earth is mineral and metal poor except for what was deposited by prehistoric asteroids and meteorites. If we could capture and return to Earth a small asteroid or meteoroid (say 1 or 2 miles in diameter), it would provide adequate minerals and metals to supply all nations for centuries to come. Now let's take a look at some of the beliefs and misconceptions related to another visitor of the night: comets.

Comets: Some facts about comets: They are composed mostly of rocks, ice, and gases. The gases are pushed away from the sun to form a tail by the *solar wind,* a stream of charged particles (plasma) shooting out from the surface of the sun at about 500 miles per hour. The tails of comets are always pointed away from the sun. The solar wind also shapes the magnetic fields of planets as the charged particles pass the planets. Comets are composed of three basic parts: the *nucleus,* which is the center made of rock and ice; the *coma,* which is composed of the gases and dust that form around the nucleus as it evaporates (the nucleus and coma for most comets are more than 10,000 miles in diameter); and the *tail,* which is made up of the gases and spreads out from the coma as described. The tails of comets can be several million miles long; sometimes comets have more than one tail. As the coma approaches the sun in its orbit, the nucleus and gases are vaporized, which increases the size of the coma and produces the tail.

Most comets originate from a theoretical reservoir of icy/gas comet material located about 50,000 to 100,000 AU from the planet Pluto. This area, referred to as the *Oort Coma Cloud*, lies well outside the solar system but revolves around the sun as a circular mass of matter left over from the formation of the solar system. Jan Hendrik Oort's (1900–1992) comet cloud theory also says that as another star or planet disturbs the orbit of the icy comet material, it sends some of it on its way to the sun as a comet. Or, if two icy chunks of comet material collide, then the path of some of this material is disrupted and starts a journey around the sun as a comet. Another theory of origin for the icy, short-term comets states that they reside in the Kuiper Belt located just outside the orbit of Pluto. The paths of comets, as well as planetoids, around the sun can be a very elongated ellipse with the sun as one focal point. Large elliptical (oblong) paths of comets are typical of those that return periodically. The path can also be shaped like a parabola, in which case the comet makes only one pass around the sun, never to return. The word *kometes* means "hairy," similar to "beard," which later was called a "tail."

Seeing a comet in the sky, night after night, was frightening for most ancient people. The motion of comets through the sky was accurately tracked by the ancient Chinese, Babylonians, and Egyptians. The Greeks held the misconception that comets were the result of the Earth's breathing luminous vapor into the sky. Both Aristotle and Ptolemy held the misconception that comets were an earthly phenomena that originated in the area of the higher air of the Earth. For many centuries ancient scientists held the misconception that comets and shooting stars were similar to a kind of lightning that were both of the Earth, and yet, above the Earth in the realm of the celestial æther. This led one to question how high the comets were? No one attempted to answer this question on a scientific basis until 1798 when two young German scientists used the *parallax method*, which is the simultaneous viewing of a comet from two different locations on Earth. It didn't work. Their conclusion was that comets revolved around the sun at a distance many times greater than the distance of the moon from the Earth because that distance could be measured with the parallax method.

The first person to make a systematic scientific study of a comet was the German astronomer Johann Müller (1436–1476), who called himself Regiomontanus. At this time telescopes had not been invented, but he could see and track the comet in its path against the stars. He kept a detailed record for a period of several nights. This took a great deal of courage because comets were considered bad omens and no one observed them in a rational manner. In 1540 a German astronomer, Petrus Apianus (1501–1552), published descriptions of several comets as fan-shaped rays through a lens and that the tails always pointed away from the sun. It was Tycho Brahe, also believing in the lens principle, who first surmised that comets were much further from the Earth than was the moon. Isaac Newton's laws of gravity finally explained that gravity was responsible for causing comets to revolve around the sun. This provided Edmund Halley the information he needed to determine that a comet observed in 1456, 1531, 1607, and 1682 was really the same comet. He was then able to predict that it would return about every 75 or 76 years, and would return again 1758 or 1759. Halley's comet made its last appearance in 1986.

There are many thousands or possibly millions of relatively small chunks of matter located in both the Oort and Kuiper belts of our outer solar system. As previously mentioned, the minor planet, Pluto, may actually be a member of the Kuiper belt of asteroids. In addition, there are billions of pieces of matter left over from the formation of the universe that were never incorporated into larger bodies such as stars and planets. Over the eons most of this matter has been attracted to larger bodies by gravity, but not all of it has been as evidenced by the existence of asteroids, meteoroids, and comets. A theory proposed in 1986 by Louis Frank of the University of Iowa is that small icy comets continuously enter the Earth's atmosphere, and that they could be the source of the water that filled the ancient oceans. These fiery snowball-like chunks might also have been the source of organic molecules that seeded life on Earth. He was ridiculed by other astronomers. Since then, astronomers at NASA have detected objects entering the Earth's outer atmosphere, verifying Frank's theory. NASA estimates that between 5 and 30 of these very small icy comets hit the upper atmosphere every minute. They break up and the ice becomes water, which may well have been the original source of water on the Earth.

Chapter 6

Conservation, Ecology, and Environmentalism

INTRODUCTION

Conservation, ecology, and environmentalism are relatively new branches of science. The early settlers of North America had a love and respect for the great nation they were exploring. As more settlers arrived they increasingly damaged the natural resources—mainly by clearing land for agriculture and over hunting animals for food in some areas. The early conservation movement grew out of concern for America's natural resources. Conservation was not thought of as a science with its own theories and principles until about 100 years ago. Ecology became a recognized science over the past 75 years. Environmental science is still developing; whether or not it is truly a science in the sense of an organized body of knowledge based on experimentation and reliable data is debatable. Environmentalism as an outgrowth of environmental science does not yet have its own theoretical base and scientific laws, but rather draws on statistical epidemiological studies and knowledge from other sciences. As such, there are many misconceptions about what environmentalism is and what affect it has on individuals and the world in general. As it develops and matures, it will be possible to observe the issues that result from the many discoveries, beliefs, mistakes, and misconceptions that now exist.

Ancient people were aware of the land and the plants and animals that provided them sustenance. For thousands of years the solution to problems related to food scarcity or other intolerable conditions was to move on and ignore any pollution or devastation of the environment left behind. Early men and women had no concept of conserving resources and thus no regard for the slaughter and extinction of the hundreds of species of animals they used for food, clothing, and shelter.

The period during which people migrated across the land bridge from Asia to Alaska and over the theoretical ice bridge from northwestern Europe to the northeastern coast of America has been revised from about 12,000 to 20,000 years ago. The long-accepted concept that early North Americans respected the Earth and its animals is now being reexamined. In *The Origins of Virtue*, Matt Ridley (1997) states that the noble savage living in harmony with nature is a myth. He says that after the arrival of early Americans, 73% of the large mammals quickly died out. Those driven to extinction by the early ancestors of Native Americans include the mammoth, giant bison, some species of bears, native wild horses, mastodons, saber-tooth tigers, giant ground sloth, and wild camels. He also states that 80% of the species of large mammals became extinct in South America about 8,000 years ago. He reports evidence that supports the idea that our noble ancestors hunted these animals to extinction for food and clothing. When early North American land dwellers depleted the land and its animals they would move on to greener pastures. This may not be as bad as it sounds. After all, the entire universe is constantly evolving. This includes the ancient natural extinction of 99% of then-existing species and the evolution of new species. It might be mentioned that humans had nothing to do with the historic massive extinction of plant and animal species—humans had not even evolved at this time in prehistoric history. In addition, all living things, including man, change their local environments through the processes of living. All living things have an effect on their environment, from simple bacteria to the most complex organisms. All life is evolving toward entropy. (The second law of thermodynamics states that disorder, randomness, and ultimate chaos [biochemical reactions and systems] become less organized.) This is a normal process of life, death, and decay.

The earliest natives of North America were most likely related to Caucasian Europeans rather than the ancestors of more recent Native Americans. Both the earliest and more recent Native Americans may have had some form of mystical, superstitious, or religious respect for animals and the Earth, but religious concern for animals does not equal conservation. Native Americans did not practice conservation. It was not until the development of stable societies based on agriculture and domestication of animals that people became somewhat aware of the need for conservation, the protection of ecological systems, and other environmental concerns. Even so, some early settlements and civilizations ignored the limitations and implications of living within a particular prescribed environment.

DEFINITIONS

In order to examine the beliefs and misconceptions related to conservation, ecology, and environmentalism it is necessary to define related concepts to develop a better understanding of the issues involved.

Conservation

Conservation involves both biology and physics. Each discipline defines conservation differently.

Biology

Preservation: This concept of conservation focuses on protecting *all* natural resources, keeping them safe from harm or destruction. This concept of conservation allows for no or very limited use of both renewable and nonrenewable environmental resources.

Wise Use: This type of conservation provides a sustainable relationship between and among animals and plants and their environment. It considers both the use and protection of our nonrenewable and renewable resources.

Most of our discussion will address the second biological definition because the first one is unobtainable for the continuation of all living organisms.

Physics

The physical law of conservation states that even if the physical properties of a closed system change, the total mass and/or energy of the system is conserved. For instance, if you have 1 gram of dry sugar and 9 grams of water, their total weight is 10 grams. If you dissolve the sugar in the water, you now have a 10-gram solution of sugar water. You no longer have crystallized sugar or pure water. But the total mass has been conserved. Similarly, Einstein's mass/energy equation, $E = mc^2$, states that there is a conservation relationship between mass, energy, and momentum. You cannot destroy the total mass/energy of a system, but you can alter the relationship of its components. There is a limited total of mass/energy in the universe.

Ecology

Ecology is the scientific study of the interrelationships of organisms and their physical, chemical, and biological environments. It is a study of systems. As a science it draws on many other disciplines, such as physics, chemistry, geology, climatology, oceanography, economics, and mathematics. Ecology also involves the study of biological communities (ecosystems). Ecology is often confused with environmentalism; the latter is often considered a political movement rather than a science.

Environmentalism

Environmentalism is often confused with ecology. Environmentalism advocates the preservation of the Earth's natural resources by society.

CONSERVATION

As previously mentioned, there are some misconceptions regarding Native Americans and North America. Douglas Preston (1997) shatters the myth that ancestors of our Native Americans were the first to settle in the New World. In an article entitled "The Lost Man" Preston writes of the discovery of a skull and skeleton on the banks of the Columbia River near Kennewick, Washington. The bones were identified as belonging to a Caucasoid man who lived over 8,000 years ago and most likely came from Europe. Later, it was determined that the Kennewick man, as he is now known, was a descendant of the Folsom man whose campsite was located in northeastern Colorado and who, in turn, was descended from the Clovis man (named for a town in New Mexico near where the relics were found). The Clovis man lived about 11,500 years ago.

It seems that some early people came across from Northwestern Europe and Greenland on ice bridges, whereas other groups of people came from Northeastern Asia across the Bering landbridge to Alaska. There is no evidence that these early Europeans or the Asian ancestors of Native Americans practiced conservation. Massive piles of bones found at the bottom of cliffs is evidence that Native Americans drove herds of buffalo off the edge, killing many, but only using the meat and hides of few. As previously mentioned, they contributed to the extinction of many species of large mammals by hunting them. The estimated native population north of the Rio Grande River before 1492 is between 500,000 to 5,000,000. No one really knows how many people lived in North America and Canada before the settlers arrived. Within a few hundred years, as European settlers moved west, the native population of North America was reduced by disease and war to less than 1,000,000. The current population of full-blooded Native Americans (excluding Eskimos and Aleuts) is just over 1,000,000 (or less than 5% of the population). Philip Shabecoff (1993) states that early Americans practiced both a hunter-gatherer and a slash-and-burn economy, at least until modern Europeans settled the continent. He mentions that when Europeans settled the New World they brought their garden concept with them.

Many of these Europeans were heedless of what they did to their environment. They had no concept of conserving when there was an abundance of just about everything. Even so, some of the explorers and early settlers developed an appreciation and love for the land itself. One misconception is that these early concerns led to the development of conservation of natural resources on a national basis. In early settlements in both New England and Virginia, local regulations were developed to limit the areas where timber could be harvested and cattle could be grazed. These regulations were the forerunners of modern environmental laws. Some early frontiersmen chafed at the restrictions placed on their freedoms.

The first conservation concepts did not come from settlers moving west to carve out a livelihood, but rather came from writers, poets, scholars, philosophers, and some scientists. Some of the literature they produced addressed the problem, or conflict, between the need to establish a new civilization, including a government that recognized laws of individual property rights, and the concept of a free forest—a place where everyone would have access to the forests and benefit from the land and nature. This literature suggested ideas about democracy that predate current debates among conservationists, business interests, those who wish to exert political control over the public and private domains, and those who do not wish such restrictions. Two misconceptions developed from this period: the belief in the myth of the noble savage as the ultimate conservationist, and the romanticized independent "woodsman." Today's current policies and practices of depending on a centralized government to regulate, control, and protect the environment and conserve natural resources is being challenged by private landowners.

Historically, the modern concept of conservation was developed about the time of the industrial revolution, when the economist Thomas Robert Malthus (1766–1834) became concerned about the decreasing death and increasing birth rates as Europe changed from an agrarian to an industrial society. In his book *An Essay on the Principle of Population* (1798, rev. ed. 1803), Malthus stated that populations increase geometrically (2,4,8,16, etc.), whereas nonrenewable natural resources do not increase and renewable resources, such as food supplies, increase by arithmetic progression (2,3,4,5, 6, etc.). Thus populations would always outstrip their resources. Of course, his big misconception was that he did not foresee the tremendous increase in renewable resources (e.g., food) through scientific developments and improved technologies. Later, in 1848, John Stuart Mill, who agreed with Malthus's principle, recognized that humans could possibly find ways to more efficiently use natural resources. The philosophies of both Malthus and Mill continue to influence people about conservation and the environment.

The term *conservation* was most likely coined by George Perkins Marsh, who probably heard it used in British India when they referred to their government's conservancies. These agencies were British commissions responsible for regulating fishing and navigation in their colonies. George Bird Grinnell, the publisher of the magazine *Forest and Streams,* was an early advocate of conservation. In 1886 he published an article advocating the protection of wild birds that were being hunted to extinction for sport and their feathers. He proposed an organization that would preserve species of wild birds. Grinnell's organization became known as the Audubon Society, which was named in honor of John James Audubon (1785–1851), an ornithologist and painter of birds.

The father of conservation in the United States was Gifford Pinchot (1865–1946), a public official in President Theodore Roosevelt's administration and governor of Pennsylvania. Pinchot was educated at both Yale and in France, where he learned about the European system for managing forests. Pinchot promoted the idea that the wise use of forests was not an isolated issue, but that wise-use conservation should be applied to all natural resources. Prior to this, people had not really considered that efficient use of all the Earth's resources could benefit all people. Two other great American conservationists were President Theodore Roosevelt and John Muir. John Muir was the founder of the environmental organization known as the Sierra Club. Both Roosevelt and Pinchot advocated the conservation concept of *wise use* rather than just the *preservation* of the nation's natural resources. In the late nineteenth and early twentieth centuries conservation was new and did not fit easily into the political spectrum, either right or left. It soon developed its own set of conflicting values and issues which have continued, and have been expanded into present day environmental issues.

As the early conservation movement grew it was very much in favor of limited use, but not necessarily complete preservation. Business supported sustained production and wise use, but not the elimination of the availability of resources required for products. Today there are proponents of both the preservation and wise-use concepts of conservation. By far, the greatest result from the early conservation movement was the application of ecology, modern science, and technology to the management of our natural resources.

ECOLOGY

The word *ecology* is derived from two Greek words: *oikos,* which means "house" or "household," and *logos,* which means "study of." In 1869 the word *ecology* was coined and defined by Ernst Haeckel (1834–1919) to mean the scientific study of how plants and animals are related to each other and their environments. It was later considered the scientific study of "natural history," a term proposed by Charles Elton. As previously mentioned, the science of ecology integrates and draws on various branches of science such as biology, chemistry, climatology, economics, geology, marine biology, mathematics, oceanography, physics, and psychology.

The idea that the impact on the environment by human activity was cumulative, and that if left alone nature was stable, was proposed by George

Perkins Marsh (1801–1882) in 1864 in his book, *Man and Nature, or, Physical Geography as Modified by Human Action.* Marsh observed human activity in many countries and came to the conclusion that wherever man existed, the harmony of nature was disturbed. He neglected to realize that *all* living things, not just humans, disturb the harmony of nature, mainly because *all* are part of the complex ecological system.

The term *ecology* is often misused by persons and organizations who are deeply concerned about the well-being of the Earth. Ecology, as well as conservation, was taught to biology majors as a specific science before World War II. Shortly after this time, however, the term "environmental studies" replaced the science of ecology as a general biology course in educational institutions. It was also about this time that the public began to confuse ecology with environmentalism. Conversely, as the terms became more interchangeable, the goals of the proponents of ecology and environmentalism became more divergent. Today, most scientists consider ecology a science based on sound theory and experimental data that is both valid and reliable. Many others consider environmentalism a political movement with a path leading to ideological issues, regulations, and other concerns. The goals of ecological scientists and environmentalists are not that far apart. Both agree than humans have harmed the environment, and both have a concern for the health of the *biosphere,* which is defined as the thin, life-supporting layer of the Earth's surface. The distinction is how the two philosophies go about searching for answers and solutions to recognized environmental problems. One group wishes to secure the best scientific data available for making public-policy decisions, whereas the other says there is just not enough time to wait for science. Some people feel that action cannot be delayed, even if adequate and reliable data is not available to guide policies.

A main branch of ecology is the study of *ecosystems*—communities of living things within a given area. The term *ecosystem* was devised by Sir Arthur George Tansley (1871–1955) in 1935. It is a rather recent concept, but was understood by ancient peoples to some degree. The hunter/gatherer societies provide an example—they just moved on when the stress they caused the local environment threatened their livelihood. Ecology is the study of a system of stresses, both natural and manmade, on local and regional habitats (land or water) by plants, animals, and humans. Ecology considers each of the parts of a system as related to the whole system. Humans cannot be separated from ecosystems any more than can other natural components of the system. This study of systems is one reason why behaviorists, economists, and mathematicians are interested in ecology. Let us consider some common misconceptions related to the idea of the interrelatedness of units to the whole community.

One misconception is that each individual plant or animal is independent and therefore has no effect on or has no relationship to other organisms, or its surroundings. Nothing could be further from the ecological truth. Every living thing affects its local environment, including other plants and animals. When a plant or animal grows, it takes in nutrients, water, and energy. When it dies, it decomposes into various chemicals and gives off energy somewhere on Earth. Thus one could say all life either pollutes or enhances the environment through the processes of birth, growth, and death. No matter how detrimental or beneficial humans may consider a particular species of plant or animal, including humans themselves, species, not individuals, are isolated entities.

The idea that a community (population) consists of a group of similar living things is another misconception. Ecologically speaking, communities are composed of an assortment of both plants and animals, which may or may not include humans. The problem is that when considering the ecology "equation," humans are often excluded. We are part of our ecosystem, like it or not. After all, the human species is a part of both the animate and inanimate ecology of the Earth. We affect other species and they affect us, as do all other things in the universe. We are part of various ecosystems, and, as such, what we do is also related to these systems.

Ecology is related to evolution in several ways. It is estimated that 99% of all species of plants and animals that have existed on the Earth are thought to be extinct. People had nothing to do with these exterminations of species because humans did not exist for most of the time life existed on Earth. It is also estimated that over 30 million different species of organisms now exist on the Earth, and that most of these will become extinct by natural evolutionary processes. Extinction and creation of altered plant and animal species on the Earth are the norm, not the exception. In terms of evolution, the Earth should exist for another 75+ billion years, when humans (*Homo sapiens sapiens*) are no longer a viable species. In the universe's total scheme of things, humans are not all that important to the ecology and welfare of the Earth.

To understand ecology one must be familiar with the following basic ecological principles:

1. Each species (plant or animal) in a community is distributed according to its own biological requirements, which usually affect other species in the community.

2. The species in a biological community exists in vertical stratification; that is, there are layers of life. For instance, some bacteria inhabit the soil. Insects, small animals, and shrubs live close to the soil, various species of trees attain different heights to fill in the layers above the shrub level. Different species of plants and animals are stratified within this community and interact at various levels.

3. The term *diversity* (variety) is related to the numbers and types of different species within a community. A community with a rich diversity has many species of plants and animals. A stable community has a balance of diverse species. It is referred to as a *climax community.*

4. Once the diversity of species is altered, an imbalance of the community may result. An example is the excess deer population in the suburban areas of the Northeast, which results from a decline in the population of the deer's natural enemies.

5. Ecologists in the United States call large, extensive communities or biological population areas *biomes,* whereas in Europe the term *biome* is reserved for plant ecology. Examples of biomes are large forest areas, deserts, grasslands, and jungles. Climate and geology have a great deal to do with determining the species of plants in a particular biome, which in turn determines the species of animals that live on these plants.

A common misconception is that an imbalance of species within its local environment or community is always bad. Often, such an imbalance offers the opportunity for new or underrepresented species to become established in the community, sometimes to the dismay of other species. An example of an unexpected consequence occurred with the importation of the Asian water lily in Jacksonville, Florida for the beautification of private ponds. The lily escaped and now clogs most southern waterways. Winter wheat is an example of a beneficial consequence of the introduction of a new species.

Sometimes the introduction of a new species is not caused by humans. For instance, very aggressive bees from Mexico and South America have invaded Texas and are breeding with native honeybees with negative results. The progeny are not good honey-makers and as they adapt they are moving north in large swarms where they have killed humans and animals.

Another misconception is that once a biological community is devastated by either natural disaster or mankind, it will be damaged forever. The restorative power of ecosystems by nature, with or without the assistance of human technology, is great. Consider forests, for example: It may take some time, but forests can reestablish themselves, particularly with assistance from humans.

Over the past century there has been a great increase in the number of trees growing in the United States. There are about 35% more trees growing in the United States today than there were 80 or 90 years ago. This increase in both the numbers and growth rates of trees provides for the absorption of some of the excess carbon dioxide produced by man and nature. (Green plants, including trees, grow more rapidly in an atmosphere rich in carbon dioxide.) The forests reestablished by both the government and private enterprise in the United States helps balance the CO_2 that at one time was absorbed by trees in areas of the world that no longer have vast forests.

To ecologists, ecosystems are the interactions of living organisms with their physical, chemical, and biological environments. Ecosystems represent the functional aspects of a biological community. They involve the interaction of factors, such as minerals, water, energy, population dynamics, and so on, with the biology of plants, animals, and humans. All are parts of a diverse community. Exploring the science behind this comprehensive approach to ecosystems is called modern ecology.

Although ecology is a relatively new science, the search for scientific information is the major distinction between ecology and environmentalism. As with any new science, new theories, hypotheses, experiments, and studies are being conducted. New concepts of systems are being explored using the new theories of complexity and chaos. As with the generation of new concepts and facts, there may be many misconceptions about ecology and the environment. These misconceptions and "facts" will be corrected over time by the use of controlled experiments and the viability and reliability processes of the scientific enterprise.

ENVIRONMENTALISM

There are many beliefs, misunderstandings, and misconceptions surrounding the three major types of environmental issues: macro, micro, and population issues. *Macro* issues are global. Some examples include the thinning of the ozone layer, acid rain, the greenhouse effect (global warming), and nuclear fallout radiation (global cooling). *Population* issues can be either worldwide or local and are based on the idea that too many people are a threat to our planet. In the long term, environmentalists consider most environmental problems to be a consequence of human over-population problems. *Micro* environmental issues are more localized, but still may have global effects. Some examples of micro issues include local water and air pollution, use of local natural resources, and industrialization.

All of these issues are too important to be left to policies, solutions, and regulations made without adequate scientific information. Let's take a look at some of the beliefs and misconceptions related to these environmental issues and the studies that provide conflicting data.

The beginning of the modern environmental movement is assumed to coincide with the publication of the popular book, *Silent Spring,* by Rachel Carson in 1962. There is no question that there was concern about environmental problems many years before *Silent Spring* hit the market. But the well-told story of a fictional town's problems with the use of artificial pesticides, particularly DDT, seemed to catch the public's interest. Many believed that this book was an attack on modern technology. Carson did not do the research related to DDT, but her book brought the issue to the attention of environmentalists. DDT stands for Dichloro Diphenyl Trichloroethane, a toxic insecticide. The degree of toxicity to humans is still an unresolved issue.

Many of the studies at that time were related to the effects of DDT on the food chain, and the fact that it reduced the thickness of eggshells, making them too fragile to hatch. Because of this effect several species of birds were placed on the endangered species list, including the bald eagle, falcon, and the brown pelican (all have since recovered). Following her book, DDT was banned from use even though it was never proved that it isn't safe for humans. It is still considered an excellent pesticide when used properly. Some studies showed that massive amounts of DDT caused cancer in animals, and other studies suspected that it might cause breast cancer in women. But at normal levels of use, none of this occurred in humans. The fact that during World War II DDT saved many lives by killing the lice that carry typhus was ignored in the debate. In addition, DDT was effective on mosquitoes, making it responsible for preventing many people worldwide from contracting malaria. The ban on DDT has resulted in an alarming increase in hemorrhagic dengue fever, a serious disease that is still a significant problem in the lower Rio Grande Valley area of the United States and Mexico. In addition, malaria has recently become a problem in many countries where DDT has been banned. Before it was banned the use of DDT greatly increased the production of plant and animal food worldwide.

In addition to individuals and their books advocating different aspects of environmentalism, a great number of environmentally oriented private organizations have been created. Their tactics for achieving their goals are to use rational scientific methods to understand and promote ecology and conservation; to use political and economic pressure to encourage both the formation of governmental regulations on the environment, and to promote a world body to enforce strict global environmental regulations, including a redistribution of technology, resources, and wealth; and finally, in some cases but not all, to resort to eco-terrorism (also known as ecosabotage, ecotage, or monkey-wrenching) to achieve their environmental goals. It is estimated that there are about 500 private environmental organizations.

According to Philip Shabecoff (1993), as the established mainstream environmental groups enter the twenty-first century, they are becoming more professional and are more willing to examine issues based on reliable science and technology. The economic and social impact of environmental regulations cannot be ignored.

There are considered to be four environmental limits (Meadows et al., 1972). They are the depletion of *nonrenewable resources* (e.g., high-grade iron ore), the decline in *renewable resources* (e.g., food supplies), the *pollution* of the Earth's biosphere, and *overpopulation*. Energy is another limit related to environmental misconceptions that may be considered both nonrenewable and renewable. The sources, production, and use of energy involve many resources, both natural and manmade. Sources such as coal, oil, and gas are usually considered nonrenewable, although we really do not know how much still exists on the Earth. Other sources such as water, solar energy, geothermal, wind, tides, and biomass (trees and other plants) are considered renewable. Broadly speaking, these energy sources also have limits.

Many people believe that gasoline is as limited as it once was many years ago. If we consider the methods used to produce gasoline 85 years ago, this would be true and we would not have enough fuel to power all the vehicles in the world. New techniques for locating, recovering, and refining crude petroleum have increased the output of gasoline per barrel of oil many times over. The same can be said for coal and natural gas. But in the long run there are limited amounts of all three of these sources of energy.

A misconception about the use of alternatives to fossil fuels to supply energy has been that society does not have the technology to harness these renewable forms of energy. This is mistaken. We *do* have the know-how. The decision not to make more use of these alternative forms of energy is often related to economics and politics—not science, technology, or the environment. There is a higher cost per unit of electricity created from renewable energy sources. This is partially a result of the advanced technology involved, as well as the lower rate of consumption of electricity created from these sources. Once the current cost per kilowatt hour increases, there will be more interest in using electricity created from renewable sources of energy.

There is one more source of energy for which many misconceptions exist: nuclear energy, which may be considered both nonrenewable and renewable. Unlike other sources of energy, nuclear energy is not dependent on the energy of the sun. It is nonrenewable in the sense that continuous fission reactions take place in radioactive elements. Thus radioactive energy sources are constantly being depleted (some at a very slow rate) and changed to other, more stable elements, while they radiate one or more forms of radioactive energy. Modern nuclear power plants, known as *breeder reactors,* can produce more nuclear fuel than is used in the process of producing electricity. Once this process becomes more feasible, most of our electricity will be produced by nuclear energy. Still the misconceptions about nuclear energy abound. Most relate to fear and ignorance about the nature of nuclear radiation. Radiation of all types has been part of the universe and has affected life from the beginning of time. True, excessive radiation, particularly accumulated harmful radiation, is a threat to living organisms, and nuclear war is always a possibility. But the belief that nuclear power plants produce and send "radiated electricity" to homes, or that radioactive vapors (steam) escape from their cooling towers are major misconceptions. The electricity produced by nuclear power plants is no different than the electricity produced by coal plants. The energy source used to produce electricity has no effect on the nature of the electricity. In addition, once we learn how to make use of spent nuclear fuel, nuclear-generated electrical power will be environmentally cleaner than coal-generated power, which produces polluting waste gases (e.g., acid rain, CO_2, and SO_2, etc.). No human lives have been lost in the United States by nuclear radiation in nuclear power plants, which cannot be said for nonnuclear electrical-power-generating plants. Another misunderstanding about energy is that it can be created. There is a finite amount of energy in the universe; the *law of conservation of energy* states that energy cannot be created or destroyed, but can be changed or converted from one form to another. For example, burning coal to produce heat, making steam that produces mechanical energy, thereby turning dynamos that produce electrical energy. Much of the energy in this system is converted to heat, not electricity. (For more misconceptions about radiation, see the section on electromagnetic and nuclear radiation in this chapter.)

Nonrenewable Resources

When viewed from the perspective of the universe, Earth is exceedingly small—just a minor speck in space. When viewed by an insect, the Earth is exceedingly large. Either perspective, small or large, implies that Earth is of a specific size, with a given diameter, volume, and surface area composed of a given amount of matter. Thus the Earth has a limited amount of everything of which it is composed. This is particularly true for Gaia, the thin biosphere in which all life exists. Even when considering the large amount of cosmic dust, comet water, and other matter that the Earth sweeps out of space every day, our resources are limited.

Since the 1930s there have been several misconceptions about the amount of the world's oil supply that have led to false predictions. In 1930 it was estimated that the world would run out of oil (and thus gasoline) in 15 to 20 years. After World War II, the estimate was revised to about the year 1970. Again and again the estimations and predictions had to be revised. A more recent estimate is that oil will last another 160 years. The cause of this misconception was that we assumed we *knew* what the petroleum reserves were throughout the world. We did not, and we still do not know exactly how much oil exists in the Earth. Another reason for the misconception was inadequate forecasting.

The misconceptions by depletionists are numerous. Predictions regarding the dwindling supply of various nonrenewable resources have been proven wrong thus far. The depletionists' predictors neglected to factor into their computer programs the changes and advances in technology related to more efficient mining, production, and use of our metals and minerals. We can use appropriate technologies that require much less of the limited nonrenewable resources. The computer programs also neglected to consider substitutions of less scarce materials for the critical nonrenewable resources. A good example of a new technology using a less critical resource is the use of nonmetallic optical fibers instead of copper wires. Communication satellites relay hundreds of millions of telephone calls, radio, and TV signals every day that would otherwise be sent on thousands of miles of copper wires. This also saves oil, and reduces the amount of air and land pollution as a result of the reduced amount of copper being smelted. Estimated reserves for many nonrenewable natural resources have increased rather than drastically decreasing as predicted.

Renewable Resources

Food

The concept of famine is based on the Malthusian model of population growth as compared to the growth of the food supply. As previously mentioned his model postulates that animal populations (including humans) increase their numbers geometrically, whereas the food supply increased arithmetically—if no other factors are considered. Unfortunately, increasing the supply of food to adequately feed an increasing human population is a challenge.

A major misconception about the food/population issue was made by Paul Ehrlich (1968) in his book, *The Population Bomb.* Based on computer models, Ehrlich claimed the Earth cannot support its human population. In 1970 he predicted that between 1980 and 1989 65 million Americans and over 4 billion people worldwide would starve. Obviously, these predictions were wrong. In 1985 he predicted that 500,000 million people would perish by starvation. When his predictions did not occur he just changed the dates or causes for new "eco-catastrophes."

Ehrlich later admitted that he neglected to consider that increasing renewable resources is possible through science and technology. In the past 50 years, research on plant genetics, improved farming techniques, economics, and systems for harvesting, storage, and transport of food products have provided an ever-increasing supply of food for the world's increasing population. Most computer population/famine-predicting programs have predicted both worldwide and localized famines for years. The big misconception here was that not all factors influencing famine were included.

Another misconception about food relates to the importance of preservation so that it can be stored for longer periods of time before consumption.

Today many nations have no way to preserve food for any length of time. Most people of the world secure, prepare, and eat their food on the same day. Food preservation is also required for the transport of food from areas of surplus to areas of need. It has been known for many decades that food can safely be irradiated with radioactive isotopes to kill harmful bacteria, insects, and microorganisms that cause disease and premature food spoilage. For instance, many of the diseases related to undercooked meat such as trichinosis in pork and salmonella in chicken can be eliminated through irradiation.

There is resistance to using nuclear radiation for preserving food because of a lack of understanding of the nature of nuclear radiation and the technology involved. The main misbelief is that the food becomes radioactive, which it does not. Several industrialized nations, including the United States, have permitted the irradiation of certain foods. This has proven very beneficial to the health of their citizens. The food industry is proposing that more foods be preserved by this method not only to provide insect- and microbe-free foods, which would provide a more healthy diet, but to enable longer storage to facilitate the export of food products to nations with food shortages.

Soil

There is no question that parts of the world, including the United States are soil poor because of poor land management and farming practices. Some soils are naturally inadequate for raising food, whereas some soils are excellent for this purpose. Environmental alarmists have predicted for decades that millions of acres of croplands will be devastated by poor farming and totally destroyed by erosion. At one time, again using an incomplete computer program, the U.S. Soil Conservation Service predicted that half of the U.S. croplands were severely eroded.

There are several problems here. The first is the commonly held misconception that when soil is eroded, it is lost. This is not exactly true. Soil does not just disappear; it is displaced by wind or water and may not be on the land of the farmer who practiced poor farming, but it did end up on someone else's land. Very little eroded soil ends up in the oceans. If it did, river deltas would extend far out to sea after many thousands of years of soil erosion. Erosion is a natural process. It is what made the mature Appalachian Mountain range lower than the younger, higher Rocky Mountains.

The second misconception is that you can't make topsoil—that it's nonrenewable. The Soil Conservation Service estimates that topsoil is regenerated at the rate of at least 5 tons per acre per year—and that is considered a low estimate by many soil scientists. In addition, crops with large, deep root systems also build soil. Legumes, which have nitrogen nodules on the roots, leave their roots behind to form humus in addition to depositing natural nitrogen fertilizer in the soil. There is also a natural nitrogen cycle that recycles nitrogen compounds. These compounds result from decaying plants, bacteria, and animal wastes that pass through the ground and back into the air. Lightning combines nitrogen and oxygen, and rain recycles these nutrients back into the soil.

Several examples of how sustainable farming techniques not only maintain the soil, but actually improve it can be found. The Pennsylvania Dutch farming areas provides one such example. For many decades the farmers' practice has been to put back as well as take from their land. Over the years, through its extension services the Department of Agriculture has provided scientific information and guidance to American farmers on soil conservation. Examples of improved farming practices are terracing, contour plowing, crop rotation, planting "green manure" crops, the periodically resting of fields, limiting use of insecticides and herbicides, increasing use of natural pesticides, eliminating flood irrigation, and more recently, using no-till planting, which eliminates the need for deep plowing. Soil is a renewable resource if used wisely.

Water

The subject of water as related to pollution will be addressed later. Here we are concerned with water as a commodity required for agriculture. One misconception concerning water usage is that water supplies are nonrenewable. The global water cycle represents the balance among the oceans and lakes, the atmosphere, and land areas through evaporation, condensation, and precipitation. There have been many misconceptions, myths, and superstitions related to the water cycle. Several superstitions from old England indicated the lack of understanding of the relationship between the water cycle and agriculture. One belief was that if you cut down or burned a fern it caused rain. There were several special days in England related to rain and agriculture. If it rained on July 15th, St. Swithun's day, it was believed that it would rain for 40 days and nights and ruin the crops. If it failed to rain on June 29th, St. Peter's Day, the apple crop would be a poor one. Rain at Easter foretold a good crop of grass and hay, and a rainy St. Paul's day, January 25th, predicted a shortage of corn and thus high prices.

Although specific results of the water cycle cannot accurately be predicted as to how much, or when, or where rain will fall, it is one of the age-old and most important natural cycles of the Earth. Again, the chaos theory is evident in predicting long-term weather conditions. There has been a recent increase in the amount of cloud cover worldwide, resulting in a commensurate increase in precipitation in many areas of the world. At the same time there are areas of reduced rainfall resulting in desertification.

One misconception is that much of the ancient water in the nation's underground aquifers is being drained for city drinking water, commercial, and irrigation purposes. This has been occurring, but again this source is renewable. In the spring and summer of 1997 the aquifer that runs under south central Texas (San Antonio area) has been refilling itself as a result of heavy rainfall in the area. In addition, techniques are being tested to "recharge" ancient aquifers. One plan is to divert excess water from rivers and streams during flooding and spring runoff of melting mountain snow into these natural underground storage areas. If successful, this will provide a renewable and self-sustaining source of water for areas that lack adequate rainfall. Not only in the United States, but worldwide the limitation of food production is affected to some degree by the availability of water. There is plenty of water on the Earth. It is just not fresh water, and the fresh water that exists is not always where it is needed most. New technologies, such as reverse osmosis, desalination of sea water, the use of treated sewage water, conservation of fresh water, and control of evaporation will continue to develop so to increase the effective use of water for agriculture and industry, as well as drinking water. Reverse osmosis, also known as plasmolysis, is just the opposite of osmosis. The osmotic process allows water, which is less dense than cell fluid, to enter the semipermeable membranes of living cells, thus providing the cell the water needed for life. Plasmolysis does just the opposite. If a hypertonic fluid of greater concentration than the fluid in the cell (e.g., salt water) surrounds the cell, water will leave the cell, it will shrink and death will occur. Reverse osmosis takes ground or river water with impurities (denser than pure water) and forces it through a membrane to produce pure water. The impurities are flushed away in the process.

The current message in the United States is that homeowners are responsible for most of the wasteful use of water in the United States, which is another misconception. True, some of us do waste water, and we are encouraged to reduce water usage by lowering the level in toilet tanks, showering rather than bathing, eliminating car washing, and so on. According to the Department of Agriculture, 90% of the fresh water used in the United States is for agriculture. Thus homeowners and retail businesses use just 10% of the fresh water. If they cut back 10% of their water usage, they would be conserving only a tiny fraction (1%) of the total used. If agriculture also cut back 10% of its fresh water usage it would result in a much greater saving. This amount of reduction in water usage by agriculture can be accomplished with more efficient methods of irrigation.

Plant Genetics

At one time people practiced genetic improvement of plants (and animals) without understanding the science involved. They learned from experience to select the best grains for seed and the most prized animals for breeding. Gregor Mendel (1822–1884), through his plant-breeding experiments, discovered the laws of natural inheritance. These laws led to the science of modern genetics, which, when applied to plant and animal breeding, results in a steady growth of our food supply. Several decades ago people believed that the United States would not be able to feed its own growing population. In a sense, applying concepts of genetics to the production of food results in an increase of a renewable resource. The production of corn in the United States increased 220%, wheat 107%, soybeans 60%, and sorghum 275% per acre over the past 50 years.

Plant breeders have developed new high-yield varieties of grains. Their work led to the "green revolution," which prevented the starvation of hundreds of millions of people. Some of these new plant varieties are smaller and therefore use less water and land space, but at the same time use more fertilizer to obtain high yields. In Japan for instance, rice production jumped from 2 tons per hectare to over 13 tons. The situation in India is somewhat astounding: Before 1970 it was a major importer of grains. After the green revolution it doubled its wheat production and by 1980 became an exporter of grains. In the meantime, India's population increased by several hundred million people.

There are some downsides to the green revolution and plant genetics. One is that it requires additional equipment, fertilizers, herbicides, insecticides, and sometimes water to assist the genetically altered plants in achieving their expected levels of production. In addition, the patents for most of these "man-made" seeds allow only one planting. A small farmer cannot save seeds for replanting because new seeds must be licensed each season from the patent holder. The subsistence farmer has difficulty in financing what is required to make use of these improvements. These problems were originally unforeseen and now require the resources of large corporate farms to realize the benefits of bountiful crops.

Pollution

As mentioned previously, the third limit to growth is pollution. Environmental pollution is caused by natural processes such as volcanic activity, forest fires, storms, by the growth and death of plants and animals, as well as by human activities.

There are several specific definitions of pollution that can be applied to either ecology or the environment: One deals with the materials that pollute—their nature, their production, their disposal, and so on. Another is concerned with ecological systems and the balance of nature, that is, air, water, land, living things, and the impact pollutants may have on these resources individually and collectively. A third definition addresses the economic and social aspects of pollution as well as the environmental effects. A generally accepted overall definition is that pollution of the environment is the discharge of material or energy into the three main environments (air, water, land) that causes either short-term or long-term damage to the Earth's local or global ecology. Misconceptions arise when trying to explain "causes," "short-term," "long-term," and "damage."

There is no question that the Earth itself, *all* nonliving and living things, including humans, are responsible for some pollution of the local and global environments; for example, sulfur dioxide and carbon dioxide from volcanoes, waste by-products from living things (feces and carbon dioxide), industrial waste by humans, and so forth all cause pollution. Many living things also contribute to the well-being of the biosphere (e.g., plants consuming CO_2 and replenishing O_2, providing fertilizer, etc.). It might be said that pollution caused by some species becomes a requirement for the survival of other species (e.g., animals produce carbon dioxide, which plants require). The problems related to pollution are ones of magnitude as well as type.

To some degree ancient people lived in harmony with their environment, as did all animals. The problem was that men and women were not willing to continue living like animals. The use of fire, early agriculture, animal domestication, tool making, and the development of technology enabled people to exert some control over the environment, rather than the environment exerting control over them. Granted, this caused degradation and pollution of the local environment, to which early humans paid little attention—they just moved on. It is impossible to divorce humans (or any other living organism) from the environment or the ecological equation.

There are numerous false beliefs and misconceptions related to pollution. We will address only a few representative samples of several types of pollution that are related to humans and their technology.

Air Pollution

There are both natural and man-made air pollutants within the Earth's biosphere. Historically, volcanic eruptions polluted the atmosphere to a degree that caused massive death for many plants and animals. Even more recent eruptions, though not as extensive as volcanic activity over the past 500 years, caused great clouds of dust and gases to encircle the Earth. Many great sunsets are attributed to recent volcanic action. The theoretical giant comet, meteor, or asteroid that crashed into the Earth 65 million years ago is presumed to have polluted the atmosphere to the extent that the sun was blocked out for several years. This killed much of the plant growth of that time, and thus, many species of animals, including the dinosaurs.

There is no argument that man has polluted the Earth's atmosphere. But there are arguments and disagreements as to the extent of atmospheric pollution, the specific causes of different types of air pollution, the seriousness of the pollution to Earth and its inhabitants, and what to do about it. From the beginning of the industrial revolution in England and later in the United States, our methods of production and energy use were not only wasteful but greatly polluting to the air, water, and land. England was the first country to recognize the effects of pollution on the people living in cities. In 1848 Sir Edwin Chadwick, considered by some to be the father of public health, caused laws to be passed that were designed to reduce air and water pollution in London and other cities. From the middle of the nineteenth century until after the 1950s air and other pollution was on the increase. But since the 1960s the rate of atmospheric pollution has been reduced, with the exception of a few gases. Human activity is responsible for most but not all atmospheric pollution. The burning of fossil fuels (coal, oil, and gas) produces primary pollutants that enter the air directly. Secondary air pollutants such as acid rain, smog (nitrogen oxides), and ozone are caused by chemical reactions of the primary pollutants

Table 6.1 lists some of the types of air pollution, their sources, and suspected effects on humans.

Table 6.1
SOURCES AND EFFECTS OF AIR POLLUTION

Pollutants	Sources	Effects
Carbon dioxide	Decay, oceans, respiration, combustion of wood and fuel	High concentrations cause death, increase plant growth
Carbon monoxide	Automobile emissions, fires, incomplete combustion	Deprives blood of oxygen, deadly at high concentrations
Sulfur dioxide	Volcanoes, decay of living matter industry, oil and paper production, chemical plants, furnaces, etc.	Causes smog, throat and eye irritations, aggravates respiratory diseases
Nitrogen oxides	Bacterial activity, lightning, acid rain, automobiles, fires	May cause eye, throat, and lung irritations in children and elderly
Methane	Termites, cows, deer, sheep, and other ruminants, biomass decay	High concentrations are toxic. May affect ozone levels
Ozone	Natural in troposphere and electric discharge (lightning); some emitted by auto and industry	Poisonous at high concentrations, irritate nose and throat, aggravates asthma. Protects from UV
Particulates	Forest fires, volcanoes, wind, car exhaust, construction debris, dust	Causes breathing difficulties for those with lung/heart problems
Other	Toxic particles and organisms; benzene, asbestos, arsenic, chloroform, viruses, bacteria, etc.	Range from mild irritation to death by lung disease and possibly cancers.

Acid Rain

Acid rain is a form of air pollution that is the subject of much misunderstanding and controversy. It is blamed for environmental damage in both local areas of the northeastern United States and southeastern Canada, as well as in England, Europe, and Africa. Acid rain is both a natural and manmade condition in which gaseous sulfur compounds and nitrogen dioxides combine with atmospheric moisture, which is then released as rain, snow, hail, or fog. One of the controversies is just how much of the acid rain is caused by natural pollution, such as sulfur from volcanos, forest fires, and the natural nitrogen in the air; and how much is caused by the burning of coal, hydrocarbon fuels (automobile exhausts), and other man-made sources.

The pH scale is used to measure acidity or alkalinity. It ranges from 1 to 14, with 7 being neutral (neither acidic or basic); a pH below 7 is acidic. Natural clean rain measures about 5 or the pH scale, which is on the acidic side, whereas rain considered polluted by man-made sources measures 4 on the scale, which is only slightly more acidic than natural rain. The problem is that after a volcanic eruption, rain downwind of the volcano can measure a 3 or 4, which may account for much of the pollution. The U.S. government conducted a $537 million, 10-year study called the National Acid Precipitation Assessment Project (NAPAP) to determine the effect of acid rain on the natural environment (e.g., lakes, forests, and crops in the northeastern United States and southeastern Canada. The scientific research proved that acid rain was an environmental nuisance, as measured by pH instruments, and may cause some increase in the natural acidity in the water of lakes located in northern pine forests. It was concluded that acid rain should be controlled in the future, but it was in no way the environmental catastrophe claimed by environmental activists, who indicated that acid rain was killing lakes, fish, trees, and plants.

One of the misconceptions relates to the acidity of lakes in the Northeast. It was discovered that most lakes that were acidic maintained this condition for hundreds of years because of the type of trees, plants, and soil in the region. Another misconception was related to the dying forests in some of these areas. There is general agreement that an excess of acid rain can be harmful, but it can also be very beneficial for some plants. In the Northeast, oak trees, balsam firs, several varieties of spruce, hemlocks, blueberries, mountain laurel, many woodland wild flowers, and rhododendrons all depend on an acidic soil, and they thrive on so-called acid rain. The NAPAP studies found that the nitrogen and sulfur compounds in acid rain act as nutrients that promote plant growth. In other words it is more beneficial than harmful to plants native to this area. Acid rain actually fertilizes millions of acres of eastern forests in the United States and Canada.

Many environmentalists believed that acid rain caused the deaths of many forests in this area. Only the red spruce was undergoing environmental stress, and this condition was caused by wind and cold, in addition to the possible effects of acid rain. The U.S. Forest Service reported that New England's forests are among the healthiest in the nation. It was also demonstrated that most of the damage to the trees people saw dying was caused by insects, (e.g., gypsy moths and pine bores), and natural effects of cold and wind—not acid rain. When Dr. James Mahoney, the director of the NAPAP study, was asked what would happen in the next 50 years if no action was taken to reduce acid rain, he is reported to have replied, "Nothing."

Congress and the Administration ignored the NAPAP report, and the work of 700 top scientists in the United States. The Clean Air Act was passed, which neglected to recognize the scientific findings of many studies, including the NAPAP report. A major conclusion of the report was that in the long term some pollution controls should be taken, but that acid rain is not the environmental crisis claimed by some scientists and environmentalists. The original Clean Air Act of 1967 and 1970, followed by the 1990 and 1997 amendments, reduced the emissions of sulfur dioxide by about 40%, and also reduced the emissions of nitrogen dioxide over 10%. This was accomplished by installing expensive scrubbers in smokestacks to remove the sulfur from high-sulfur oil and coal, and by installing catalytic converters to remove nitrogen oxides from automobile exhausts. Actions are being taken to correct environmental air-pollution problems, but often without adequate scientific studies being conducted. When available, the results of reliable studies are often being ignored by those who make public policy.

Particulates

Air pollution is related to the effects of particulates, which are small particles like smoke, soot, fine dust, volcanic ash, and so on that are light enough to become suspended in the upper atmosphere. There are several misconceptions about the effects of atmospheric particulates. One is that these "pollutants" cause global warming or a greenhouse effect. Just the opposite has been observed by satellites, which determined that solar radiation is reflected from particulates in the upper atmosphere. These particulates were considered to be one of the causes of the global cooling period from about the late 1930s to the mid-1970s.

In February 1997 the Environmental Protection Agency (EPA) reported on several selective studies on particulate matter of 2.5 microns or less (soot is 10 microns in size) in the atmosphere. The misconception is that particles of this size are dangerous to the well-being of humans. Scientific studies showed that there was no association between this particulate matter and ill heath, including studies conducted by the EPA. Particles the size of 2.5

microns are so small that they exists in clean ambient air; that is, the air we breathe when visiting the countryside. For instance, the air above pristine pine forests, with no industry or heavy automobile traffic nearby, "breathes" volatile particles into the air. Trees in pine forests produce particles of 2.5 microns in size and larger. Some smogs are the products of a healthy forest. This is what causes the "smokey" haze above pristine pine forests (e.g., the Great Smoky Mountains).

The health rationale for further controls to reduce 2.5-micron particulates from the air misrepresented the scientific data, indicating that it would protect the public's health, particularly preventing asthma in children. This is a misconception because children's asthma is caused by indoor air and allergens (mites, cockroaches, dust, animal dander, molds, and pollen)—not outside air. Asthma is aggravated, but not caused by, high levels of surface ozone and air pollution. To reduce the level of particulates to 2.5 microns will not substantially improve the air, but would substantially increase the cost of pollution controls.

Many misconceptions about the environment and its effects on our health arise from our lack of understanding of the nature of epidemiological studies. Almost all studies that try to determine the relationships of a disease, its causes, and its effects on people are done by statistical methods derived from events. Epidemiology is not an exact science. It is based on observations of groups and subgroups of people, not on hard cause-and-effect relationships derived from experimental research. The follow-up procedures and statistical treatment of data are used by epidemiologists and public-health professionals for the protection of the general public. Until another system is developed, epidemiological studies are the most effective method of tracking the spread of diseases.

Overpopulation

The fourth environmental limit is overpopulation. The world's population increased from fewer than 1 million people in prehistoric times to over 6 billion at the beginning of the millennium. It is predicted to reach 8 to 14 billion before the year 2100. There are numerous misconceptions related to population increase and limits. As previously mentioned, the current advocates of the Malthusian doctrine neglect to consider the rapid increase in the growth of the food supply, increases in efficient use of other resources, and the developments from research and technology. Unlike animals, whose populations may be limited by the local food supply, intelligent humans have alternatives. The reasons for the increase in the world's population are varied. The birth rate is rising faster than the death rate, partially a result of an increase in life expectancy. People live longer because of better nutrition (less starvation), improved medical and health care, and a general increase in the standard of living, not only in some of the developed countries, but worldwide. There are many other social, political, religious, and economic factors that impinge on the rate of decrease or increase of the human population. And, of course, as an evolutionary necessity there is the natural urge for all living things to reproduce.

The concept that population density or overcrowding is a worldwide problem is a misconception. In 1997 there were 97 people per square mile worldwide. By the time the world's population reaches 10 billion, there will be 192 people per square mile, on the average. There are misconceptions related to population density. In some regions, high population density is no problem, whereas in others it causes severe environmental problems.

The misconception concerning overpopulation is that overcrowding (population density) is the cause of environmental and other problems. The reverse is often the case when considering the effects of population density on the environment, the health of the citizens, and other problems for many of the low-density nations (e.g., the degradation of the Amazon Rain Forest). On the other hand, Indonesia, with a high population density, is devastating its forests. Several European countries with relatively high population densities have some of the highest life expectancies, as well as the fewest environmental problems in the world, and some of the best managed forests. There *is* some limit (yet unknown) to the number of people the world can support—if a high standard of living is to be maintained. If, for some reason, the world's population is forced to drastically reduce its standards of living to the point that we all live like prehistoric humans, the world's population would be reduced drastically, with a concomitant improvement in the environment. Statistically, the industrialized nations with the highest standards of living also have the lowest birth rates. The United States requires immigration to maintain a slight increase in population. At the same time it is one of the most environmentally regulated and ecologically aware nations in the world.

Wastes

Humans produce two major forms of waste: biodegradable and non-biodegradable. *Biodegradable* waste, such as feces and garbage, breaks down over time with the aid of microorganisms. *Nonbiodegradable* waste, such as old cars and carbon dioxide, does not decompose by biological means. Let's take a look at some of the beliefs and misconceptions related to each of these classifications.

Biodegradable Waste

Even today many people dispose of human wastes (urine and feces) as did our ancient ancestors. They either just relieve themselves when needed and move on, bury the waste, or use it for fertilizer. The outhouse was an improvement over the World War I "latrine trench," but not much. In large cities the disposal of rain runoff, human waste, and garbage became a major problem as city populations increased. In the 1600s the practice was to throw the garbage into designated areas of the streets to later be picked up by rakers, who usually dumped the trash into the sea. Human wastes were collected from the homes by "night-soil" men who then sold it for fertilizer. The cesspool, followed by the septic tank, were additional improvements.

The water closet was first developed in about 1550 by Sir John Harington, an inventor in Yorkshire, England. It was later improved and patented by Joseph Bramah in 1778. His toilet had a valve that could fill a tank, the water and waste was then dumped into a basin that had a trap at the bottom that prevented fumes from backing up. It worked much like today's toilets. (About 100 years later, Thomas Crapper developed and manufactured the modern toilet now used in the United States.) But in the days before sewage disposal plants there was still a problem because a cistern or cesspool was usually dug beneath houses into which the waste from these new toilets was dumped. This proved inadequate as the cesspools leaked into the water supply in areas of high population density, promoting the spread of disease. A system of sewers or tunnels leading to the Thames river was developed, which many people thought would solve the human-sewage problem.

Over the years, different types of sewage-disposal systems were developed. Contrary to expectations, most proved inadequate to handle the human wastes from large population areas.

The waste input to modern sewage-treatment plants is more than 98% water, which causes some of the problems of disposal. Most people have a misconception as to how much sewage water an average household sends to the treatment plants. Each European household is estimated to produce over 100 gallons of waste water per day; the American household produces over 150 gallons per day (Walisiewicz, 1995, p. 130). In a city with several million people this amounts to a tremendous amount of waste water to be treated. There are three main processing stages: The primary and secondary stages treat the sewage for bacteria and remove sludge, particulate matter, and odors. These two steps provide waste water that is not exactly pure, but can be further aerated and treated for agricultural use. A third stage, called tertiary treatment, is sometimes used, providing an effluent that is suitable for recycling into streams and city water supply systems. Additional techniques have been developed for manned spacecraft in which human wastewater is recycled into drinking water. These advanced sewage-disposal systems, combined with improved drinking water treatment plants, are expensive and are not used in most countries of the world. Therefore safe disposal of human wastes for an ever-growing world population will continue to be a problem for future generations.

There are several misconceptions related to garbage and other biodegradable trash. Prehistoric humans maintained a garbage dump called "kitchen middens," which have provided anthropologists with much information about ancient cultures. Humans have always found some place to dump their refuse. The problems are compounded by an increase in world population, the concentrations of people in cities, and a society that overpackages the products it consumes. The town garbage dump with its smoldering stench still exists in many countries.

Cities use three main methods for dealing with wastes: they bury it, burn it, or recycle it. After World War II landfills replaced garbage dumps. The concept of a landfill is viable for most nontoxic and some toxic wastes.

Most people believe that homes produce most of the solid waste in the United States. This is a major misconception. Half the solid waste produced by modern societies comes from agriculture (Walisiewicz, 1995, p. 428). Not all of this goes into landfills. Much is animal manure used as fertilizer. Plant residue is recycled into humus and as soil stabilizers. About 40% of all waste comes from metal processing and mine tailings. About 5% is industrial waste, which if not toxic is buried. Only 5% of all waste produced in the United States comes from household and commercial (office and retail) refuse. This remaining 5% attracts the most public attention, because it impinges on our lives most directly, and we have most control over its fate.

Much unusable land has been reclaimed for landfills. A well-run landfill has a liner on the bottom to prevent seepage, a liquid recovery system, and a bulldozer to compact the daily deposits and to spread a thin layer of dirt over the waste. Because of the increasing cost of acquiring land, strict regulations about lining the bottom of the landfill to prevent seepage, and a great increase in hauling costs, some people are using back country roads and out-of-the-way areas to dump their garbage.

Biodegradable waste is also disposed of through incineration. The misconception that burning trash pollutes the environment still persists. True, old methods were inadequate and did cause air pollution. New technology has developed superhot incinerators (pyrolysis) that completely consumes the garbage, producing a minimum of ash and gases. This process also produces excess heat that can be used to generate steam for heating cities or to generate electricity. Pyrolysis combustion is one of the most advanced and safest disposal systems for any waste. This is a form of high temperature destructive distillation that chemically decomposes many different types of waste, including toxic wastes. The resulting steam and gases are mostly methane, hydrogen, carbon monoxide, and carbon dioxide, plus ash. This process reduces the total volume of the waste and trash by about 75%, with about 20% remaining as ash, which can be buried in a landfill, and 5% as "fly ash" and gases, which are either recovered or removed by scrubbers. This process requires prodigious amounts of trash in order to keep the system operating at maximum efficiency. Therefore it is only used in large cities with enormous waste-disposal problems. As mentioned, the excess heat produced can be sold to make the operation more cost-effective.

Another method of disposal is to convert wet garbage into humus for use as fertilizer on crops. Another converts garbage into methane or alcohol for use as fuel.

There is also a misconception about the safe disposal of toxic wastes. There are many classifications of toxic wastes; for example, medical/biological, chemical, flammable, radioactive, and so on. They can be handled and safely disposed of in several ways. One possible solution is resource recovery, which is to find a use for these waste substances and reuse them for purposes other than their original intent. Another method is to bury the waste in specially designed landfills. And the third, which is safe for all types of toxic materials, is *pyrolysis*—high-temperature burning. Currently, the most environmentally safe methods of toxic-waste disposal are burying or incinerating the final end products.

Nonbiodegradable Waste

In general, nonbiodegradable wastes are substances that are not broken down by bacteria and other biological or natural chemical processes (i.e., metals, glass, ceramics, mining wastes such as runoff and slag, and some chemicals). Some nonbiodegradable materials can be reused and recycled if it is economically feasible to do so, if not, they can be buried or incinerated. This brings up a number of current misconceptions related to recycling.

Recycling: Recycling is not a new concept. Ancient people melted down their metal tools and weapons to recast new ones. When furniture or other wooden objects were no longer useable, they were used for fuel.

There are many ways to treat recycled material to prepare it for further use. Some examples are magnetic separation of metals, shredding, screening (for size), separating by weight, washing, and manual separation of the material. Some systems grind up the material and add water to make a slurry, which is then run through magnets and centrifuges to separate different substances by density. The heavy metals sink, the lighter materials float, and the centrifuge acts like a cream separator. If the reclaimed material contains unusable substances these are burned or buried.

It takes an enormous amount of recyclable material to keep a large plant operating. There are many misconceptions related to recycling. To appreciate what is involved, we will consider the example of paper recycling. There are at least two major misconceptions related to the recycling of paper. First, many environmentalists believe that recycling paper will save trees, and second, they believe that it will curb the need for future landfills. Both are wrong. The best timber is used for lumber, not paper pulp, so paper production does not affect the national forest. In addition, numerous public and private reforestation programs as well as tree farms plant and harvest crops specifically for paper pulp. In the United States more trees are planted each year than are cut down; these new trees are planted mostly by the paper and timber industries. The paper industry is one of the reasons that we now have 35% more trees in the United States than we did many decades ago. That in itself might be considered a form of recycling.

Another misconception about recycling in general and recycled paper in particular is that it is cheaper to recycle old newspapers to produce pulp for new newspapers. Actually, the increased cost of handling and processing the waste paper makes recycling more expensive than starting with plant fiber. The increased cost of recycled bags and paper products is passed on to the consumer. One suggestion to reduce the cost of recycling paper is to find ways to reuse old paper in different forms. Instead of recycling used paper into pulp, it can be shredded and used as packing material or, if fireproofed, used as insulation. It can also be converted, along with wood scraps, into fireplace logs. Probably the most cost effective use of recycled paper is to sell it to the fiber-poor countries of the world. Asia, Central and South America, and Mexico in particular need large amounts of fiber that they cannot produce themselves. There are other renewable plant crops besides trees being considered for a less expensive source of paper pulp, but they have not yet become viable, worldwide sources. One promising new source of fiber for paper pulp is an old plant, kenaf, now being grown for a the newspaper pulp industry in southeastern Texas.

The second misconception has to do with landfills. The disposal of newspaper in landfills accounts for about 5% of the total, another 2% of landfill totals is made up of paper products such as bags and stationery. So if we eliminated all paper products from landfills, 93% of the landfill would still be needed (Krulak, 1993). A better use would be to package used paper and sell it to countries that have destroyed their forests and need fiber. The popular myth is that the United States is running out of land that can be used for waste disposal. According to economics professor Roy E. Cordato (1996) there is no landfill shortage. Cordato states that if all the solid waste for the next thousand years were put into a single space, it would take up 44 miles of landfill, a mere 0.01% of the U.S. land. Every person in the United States generates about 20 times his/her own weight in garbage each year. This is almost 350 million tons of trash and garbage produced annually. Only a small portion of this is recycled, the remainder is disposed of in landfills or incinerated.

Another misconception is that products made of recycled materials (plastics, metal, paper, etc.) can continue to be recycled over and over again.

Most recycled materials, such as plastic bottles, have a limit to the number of times they can be recycled before eventually ending up in a landfill. No matter how often something is recycled, sooner or later it ends up in a landfill or is incinerated. Recycling delays disposal problems.

Another misconception is that recycling saves money as well as resources. This belief ignores the economics of supply and demand. There are two products that are less expensive to produce from reclaimed/recycled material than when produced from original raw materials: aluminum cans and plastic bottles. These two examples are the exception. When the total process is considered, recycling of most materials is not cost-effective. Many cities that have built large recycling plants find that they are very expensive both to build and operate. It is the citizens who pay for this ill-informed policy of recycling through increased taxes. Recycling is not used much in other countries. Germany recycles about 15% of its waste, whereas the United States recycles only about 11% or 12% of its total waste. Some underdeveloped countries recycle only in the sense that they repair and reuse original items to a much greater degree than do developed countries.

Recycling is not always the best way to use and reuse resources, but it gives the public a feeling that it is doing something useful. The concept of complete recycling is a misconception that is impossible and unnecessary. Incineration and composting [landfills] were once accepted by cities and counties as companions to recycling—until pressure from environmentalists caused politicians to abandon them almost completely in favor of total recycling programs.

OTHER ENVIRONMENTAL ISSUES

Scientists are concerned about how the public is informed and misinformed about environmental issues related to the ozone hole, global warming, toxic substances, radiation, biotechnology, and so forth. The dissemination of environmental information that has not been adequately researched by scientists can have a devastating effect on our economy and society in general, and may or may not improve the environment. Let's take a look at several additional ecological and environmental issues for which there are many beliefs and misconceptions presented by both environmentalists and ecologists.

Ozone

Ozone is not just a man-made pollutant. It is one of the three natural forms of oxygen. Atomic oxygen (O) is sometimes referred to as nascent or newborn oxygen. Diatomic oxygen (O_2) is the molecular form of the gas that sustains life. Triatomic oxygen (O_3) is known as ozone. Small amounts of ozone are produced at ground level by lightning, electrical discharge, exhaust from automobiles, landfills, emissions from power plants, chemical solvents, and the reaction of volatile organic compounds in the presence of sunlight in hot weather. It exists on the surface of the Earth in low concentrations and can be detected by its sharp odor, which has a distinctive pungent character. It is poisonous in high concentrations, and it contributes to the formation of some types of smog. Ozone is an excellent oxidizing agent, which makes it useful for bleaching. It is also used to sterilize air and water. The ozone that has become an issue is produced by a natural photochemical reaction in an area of the lower troposphere known as the *ozonosphere,* which is about 15 miles above the surface of the Earth. This ozone layer is normally very thin relative to the entire depth of our atmosphere. This thin layer of ozone forms a partial protective shield that prevents most of the ultraviolet (UV) radiation from reaching the Earth's surface. Excessive exposure to high-energy radiation, including UV radiation, can affect crop growth and increase skin cancer in humans. In very high concentration, shortwave-length/high-energy radiation may cause genetic mutations in plant and animal cells.

The natural photochemical reaction that takes place in the ozone layer is reversible. O_3 is broken down by ultraviolet light from the sun to O_2 and O, which in turn recombines to form O_3. The current issue surrounding ozone develops when this balance is interrupted by both natural and human activity. The majority of the ozone is produced in the region above the equator (area of greatest sunlight) and slowly drifts toward each polar region. Periodically the polar layers of ozone become thin; holes can appear at particular times of the year. The holes become larger in the autumn and smaller in the spring. One problem is that records do not go back far enough to demonstrate whether these holes are part of a natural cycle or are related to other natural causes such as changes in the sun. The holes are largest between September and November. They have been measured over Antarctica every October since 1979 and show both an increase and a decrease in size. The eruption of several volcanoes over the past few decades has spewed enormous amounts of sulfuric acid into the atmosphere, which adds to the destruction of the ozone layer. This type of natural enlargement of the hole and its natural recovery has been going on for hundreds of thousands of years. There is still no evidence as to what effects this expansion and contraction will have on plants and animals.

Man-made chemicals, particularly chlorofluorocarbons (CFCs), chlorine compounds, sulfur oxides, and halons (fluorine compounds) do affect the ozone layer. CFC-11, CFC-12, and related chemicals were used in refrigeration, air conditioners, propellants in spray cans and fire extinguishes, and even in asthma inhalers. As far back as the late 1960s and 1970s laboratory studies conducted by scientists at Du Pont demonstrated that UV could split the CFC molecule to release chlorine atoms, which in turn could break down the O_3 to form ClO (chloromonoxide) and leave an O_2 molecule. This stealing of a nascent atom (O) of oxygen from ozone is repeated as in a catalytic reaction. In theory, this meant that the single O atom was not available to recombine with the O_2 to reform O_3.

There is some question whether this same laboratory reaction really takes place in the ozone layer, or if there are other circumstances in which nature plays a part in reducing the ozone layer; it is possible that there are other compensating factors for the natural production and destruction of ozone molecules. A fact often overlooked is that a reduction of ozone molecules is not only dependent on chlorine molecules in the CFCs (as well as from the oceans), but that a very cold temperature (–80°F or lower) is also required in the upper atmosphere to complete the photochemical reaction. Even so, scientists predicted a decrease, followed by an increase, in the level of ozone, particularly above the Earth's polar regions. One of the lowest readings of the ozone was in December of 1994. Since then, the size of the hole over Antarctica has decreased. Regardless, the United States and other countries have agreed to limit and eventually eliminate the production and use of chemicals suspected of affecting the ozone layer. Recent reports indicate a worldwide reduction of the production and use of CFCs, followed by a slow but steady improvement in the ozone layer.

Ozone located near the surface of the Earth comes mainly from the reaction of sunlight on the nitrogen oxides emitted from automobile exhausts, as well as from industrial and natural sources. There is a misconception that this ground-level ozone is spread evenly over the world. High levels of ozone that can cause respiratory distress are usually found in areas of congested traffic that are surrounded by mountains, such as Los Angeles, California. Geographic areas that form valleys, have a sunny climate and inadequate air flow, and often experience air-temperature inversion that prevents ozone from dissipating. There are a number of meteorological and geographical factors that cause air pollution to be greater in some locations than in other.

The ozone layer is about twice as thick over the poles as over the equator—thus more UV reaches the Earth at the tropical and subtropical regions. Ozone is produced over the equatorial regions of the atmosphere, where it is thinnest, and moves toward the polar regions. The intensity of UV increases as altitude increases. Therefore people living in Denver receive much more UV than do people living in a northeastern city.

Another misconception is that increased UV from a thinner ozone layer is harmful to all plants and threatens the world's food crops. Leading scientists conducted experiments on the effects of UV on plants in which the UV level was similar to what might occur with a 16% to 25% decrease in the ozone level (which is higher than any level that has ever occurred). Over 100 different varieties of soybeans, corn, rice, and wheat were tested. They found that about 40% of the species were tolerant or unaffected by the increase in UV, which indicates that there are species of plants that are able to withstand levels of UV much higher than those experienced by any hole in the ozone layer. The scientists also discovered that some species of soybeans are adversely affected by the UV, whereas others are improved by the increase. Their conclusion was that plants are not a good biological measure of the effects of increased UV levels as some are adversely affected and some benefit from very high levels of UV. They also concluded that a small decline in the ozone layer and a small increase in UV is no threat to the world's food supply.

Another ultraviolet radiation misconception was the possible effect of UV leaking through the ozone hole over Antarctica. Environmentalists claimed that the polar ecosystem, including phytoplankton, fish, and animals were being destroyed. Scientists have determined that the ecosystem of Antarctica is not on the verge of collapse due to an increase in UV in the polar summer. There is no demonstrated global damage to plant or animal life resulting from the ozone hole.

Some believe that there is an inverse relationship between the levels of ozone and UV. As the measured ozone level has declined globally, scientists have not measured a similar increase in ultraviolet radiation, particularly in the northern hemisphere. There is even some indication that as the measured levels of ozone decreased, so did the levels of UV!

Other misconceptions relate ozone depletion to the concern over the greenhouse effect and global warming.

Global Warming

Several decades ago environmentalists were concerned about the possibility of a nuclear war that would block the sun and thus cause a "nuclear winter" and result in a new ice age that would end in death to all living things. There were also predictions that industrial gases and other air pollution would cause a cooling of the Earth, which could be harmful to plants and animals. Neither has happened. More recently the alarm has been over an increase in global temperatures during the past 100 years of less than 1°F.

The global warming scare is rather recent. In 1988, James Hansen, of the Goddard Institute for Space Studies, reported to a U.S. Senate Committee that he was sure that the warmer weather of the 1980s was due to the greenhouse effect. This was picked up by the media and environmental activists as a new cause. Meteorologists, climatologists, and other scientists were not yet convinced of the extent of global warming and insisted on more research.

The greenhouse effect is real, but misunderstood. A glass-enclosed green-house used to grow flowers and vegetables is designed to keep the cool air and wind outside, and the warm moist air inside. The Earth's greenhouse effect is somewhat different. If we had no atmosphere the average temperature of the Earth would be about 30° to 40° below 0. During the daylight hours most of the sun's heat is reflected back into space by the Earth's atmosphere (including clouds and dust), but some of the sun's heat is contained and stored in the Earth, water, and gases of the lower atmosphere. A cloud cover at night acts as an "insulating blanket," which keeps the heat of the Earth and the lower atmosphere from radiating into the outer atmosphere and on into space. This is why cloudy nights are usually warmer than clear nights.

In addition, an increase in cloud cover also reflects the sun and causes a slight cooling of the Earth. The Earth's cloud cover has been increasing over the past several decades because of an increase in moisture. Scientists say that there is a *net cooling effect* caused by increasing global cloud cover, not a warm-ing effect, because of the increased reflection of sunlight by the clouds. In other words, clouds help keep the Earth warm at night, but during the day they reflect more heat than is retained at night. The chemicals and gasses that make up the atmosphere vary greatly. The amount of water vapor in the atmosphere has an effect on its temperature, and thus has some relationship to the greenhouse effect. Other gases in the atmosphere, both natural and induced by man, assist in maintaining the warmth of the Earth. Over the past billion years the Earth's temperature has not varied more than about 15°C due to the emission of gasses produced by biological sources (bacteria, etc.), and the evaporation of water into the atmosphere. Simple microorganisms produce about 0.5×10^9 tons of nitrous oxide (N_2O), and about 2×10^9 tons of biogenic methane (CH_4), which account for much of the warming of the atmosphere. Particulate matter (very small air-borne dust, smoke, soot, vol-canic ash, etc.), as well as other man-made causes may add to the effects of the natural gasses responsible for the greenhouse effect. There is some evi-dence that particulate matter of about 5 to 10 microns in size (smoke, soot-size carbon, dust-size particles, and ash from volcanoes) in the upper atmo-sphere may cause global cooling rather than warming by reflecting a portion of the sun's heat. There is some indication that changes in the sun's output of energy may have some effect on global warming and cooling.

Scientists were aware of the greenhouse effect, but it did not become a perceived problem until some environmentalists maintained that global warming was caused by industrial nations increasing the amount of carbon dioxide (CO_2) released into the atmosphere. The increase in CO_2 is the result of increased fuel consumption, mostly coal and oil used for energy, but also wood burning, combustion of other carbon-containing materials, vol-canic eruptions, as well as animal respiration, and the decay of biomass. The reduction of forests, particularly the rain forests (trees use carbon dioxide), also led to an increase in global carbon dioxide.

There is no question that carbon dioxide emissions have increased. EPA figures show that about 1.75 billion metric tons (bmt) of carbon from the burning of fuels was emitted into the atmosphere in 1950. This increased to about 5.5 bmt in 1985. This seems like a great change in the percentage of CO_2 in the atmosphere. Carbon dioxide comprises about 0.033% of the molecules that make up the atmosphere. This has increased by approximately 12% over the past 100 years, but is still low. The emission of carbon dioxide into the atmosphere continues, but has slowed down.

There are several misconceptions that arise when considering the levels of atmospheric carbon. First, only about half of the total carbon dioxide (from all sources) enters the upper atmosphere. Most of it is absorbed by the oceans to form calcium carbonates such as limestone, coral, and the shells and bones of organisms. The Earth's land plants absorb much of the rest. The second misconception is that the Earth's atmosphere is changing drastically. There have always been changes in the percentages of the reactive gases such as nitrogen (+78%) and oxygen (+20%) present in the atmosphere, but carbon dioxide has existed at about the same level (0.03%) for a billion years. For hundreds of years global temperatures have also always fluctuated naturally. There seems to be a 10,000 year "warm" period between the most recent ice ages in the Earth's history. We are now entering the last 1,000 years, or the downside, of our current warm period, which means we may be close to the onset of a new ice age.

Deforestation in some parts of the world is partially compensated for by the massive reforestation of Europe and North America. Some scientists claim that the forests in Europe and North America take up the excess carbon dioxide produced by modern industry. The forests, grasslands, crops, and plants of all kinds benefit from an increase of CO_2 in the atmosphere. One study showed that there was no measurable carbon dioxide in the air within the area from ground level up to 1 meter in height in a field of corn growing in bright sunshine. Evidently it was consumed by the process of photosynthesis within the corn leaves.

Just how much warming is the Earth experiencing with an increase in carbon dioxide, and to a lesser extent an increase in both natural and industrial methane, sulfur and nitrogen oxides, and other compounds? This depends on where the measurements are made and exactly how they are made. Back in the 1950s several scientists predicted a warming trend of several degrees as a result of the increase in carbon emissions. Not much attention was given to these findings because the Earth, at that time, was experiencing a cooling phase. The amount of preindustrial CO_2 in 1850 was about 275 ppm, today it is about 355 ppm. These figures have been used in computer programs that model global-warming effects, but they also use the *projected* figure of 600 ppm, which gives a completely arbitrary result.

Another misconception is that the atmosphere is and has been consistent over the centuries. It is estimated that the light (and heat) from the sun has increased by 25% to 75% in the last 4.6 billion years. The temperature and composition of the atmosphere has also varied over the last 3 billion years. Even so, since the beginning of life, organisms have adapted to these changing conditions.

A major misconception held by some environmentalists is that the North and South Poles will warm up several degrees, resulting in worldwide flooding of coastal cities. Actually, the temperature at the North Pole has decreased by several degrees in the last 50 years. Other scientists think we are on the verge of a new ice age, which would increase the ice at the poles and decrease ocean levels.

This flooding scare is often exploited to influence developed nations to reduce the gas emissions that may cause environmental destruction. Most industries in developed countries have been reducing atmospheric emissions for some time, but global carbon dioxide emissions and some other pollutants continue to rise as a result of emerging industries in developing countries. Many mature industries now recognize that it makes economic sense to reduce pollution.

In an editorial published in the *Valley Morning Star* of Harlingen, Texas, one of the conclusions was that it is ridiculous to think that individuals and governments have much control over the global temperature. Carbon dioxide emissions produced by man are supposedly the main cause of global warming. Yet they are dwarfed by the amounts produced by nature in the form of volcanic eruptions. The main point of the editorial was the possible effects on society and the economy caused by environmental restrictions imposed by international treaties on developed nations in favor of undeveloped nations.

Another misconception is that the Earth and its atmosphere are static, nonevolving entities. According to proponents of the Gaia hypothesis, the Earth and its atmosphere, the biosphere in particular, are a homeostatic system of components that is evolving. This theory contends that, historically, microorganisms (bacteria) are the main source of the gases in the atmosphere, and were responsible for the evolution of life forms. After all, humans have been on the Earth for only a few thousand years, whereas gas and waste producing bacteria have been around for thousands of millions of years. Therefore some scientists claim that as humans are only 1 of over 30 million species, their impact on the atmosphere may be overestimated.

Toxic Substances

Many of the beliefs and misconceptions related to toxic substances stem from a lack of understanding of the nature of the substances themselves, uncertainty as to how toxicity is determined, and what effects these substances have on plants, animals, and human beings. Public-health officials consider three main factors related to toxins in the environment: cancer, birth defects, and miscarriages. The exact causes of these occurrences are not known. Environmental activists believe that all three are caused by toxic substances, or that occurrences have increased since the advent of man-made toxic substances. People contracted cancer, gave birth to infants with defects, and had miscarriages long before man-made toxins existed, however. This indicates that these health problems have been with humans always. Increases in these illnesses may be attributed to improved reporting of diseases and longer lifespans.

Cancer Scares

There are thousand of things both natural and synthetic that may cause cancer. These include viruses, cosmic rays, ultraviolet and other forms of natural radiation, natural pesticides produced by plants, and some man-made chemicals.

Scientists have a good idea of what causes some types of cancer, but the mechanisms of cancer within cells are not clear. When increases in cancer are reported, a factor often ignored is that it is mostly a disease of older people. Because life expectancy has increased to about 80 years of age, longevity provides an explainable increase in the number of cancers. Men and women between the ages of 18 and 40 are least likely to develop cancer, but the ones they are prone to such as testicular and ovarian cancer, have existed for a long time, as have most cancers. Once the increase in longevity is considered, there is no significant increase in the incidence of most cancers. As people live longer, they have children later in life and thus increase the incidence of miscarriages or giving birth to babies with birth defects. As people live longer, the chance of contracting cancer increases rapidly. Richard Doll and Richard Peto, epidemologists of Oxford University, have stated (1993, p. 59) that there is no convincing evidence that a general increase in cancer is related to factors associated with the modern industrial world. Any increases in cancers (as well as many other diseases) are not only caused by aging, but also result from lifestyles that involve activities such as sun bathing, smoking, improper diet, and lack of exercise (and unprotected sex, e.g., AIDS), none of which are related to industrial or agricultural toxic substances and are preventable.

There is a misconception that organic fruits and vegetables, those raised without synthetic pesticides and herbicides, contain no toxic substances. All vegetables, including those grown organically, contain dozens of natural toxins. These toxins are the natural pesticides that plants have developed over eons of time to protect them from insects. One leaf of cabbage contains about 50 natural pesticides. Examples of fruits and vegetables that contain high levels of *natural* carcinogens are apples, bananas, broccoli, cabbage, carrots, brussels sprouts, celery, grapefruit and orange juice, honeydew melons, mushrooms, peaches, black pepper and other spices, raspberries, and turnips. Of all the natural toxic substance found in a variety of fruits and vegetables that have been tested in animals, about 75% have proven carcinogenic. This percentage is similar to that found in synthetic pesticides and other chemicals.

A relative risk index called HERP (human exposure dose/rodent potency dose), which compares the maximum rodent dosage to the minimum human dosage, was developed for measuring the degree of carcinogenesis of various toxic substances. This index provides some interesting test data. One liter of tap water is rated at 0.003 because it contains chloroform, which causes cancer in rodents. This is higher than the index for bacon. Peanut butter has an index of 0.03; diet colas = 0.06; mushrooms = 0.1; beer has the highest index—2.8, because it contains the carcinogen alcohol.

Several misconceptions can result from the procedure of feeding or injecting laboratory mice or rats with samples of toxic substances. Scientists use the MTD (Maximum Tolerated Dose), which is the most they can feed or inject into an animal without killing it by the dosage alone. Never is the equivalent dosage approved for humans. Here is where the problems arise: First, animals are used instead of humans and there is little evidence that what causes cancer in rats causes cancer in humans; some rats do not get cancers from the same dosage of a toxin as do their mice cousins. In one study, over 25% of carcinogens tested did not produce cancers in rats, but they did in mice, and vice versa. Rats are not mice, and neither are humans. Many man-made pesticides exist in barely measurable amounts (several parts per billion [ppb] or parts per trillion [ppt]) in food plants. At these levels, synthetic pesticides are found much less often than naturally occurring carcinogens. At these extremely low levels, there are no risks to either mice or men.

A similar misconception is related to the *no-threshold theory*, which states that there is no safe dose for humans if a massive dose of a synthetic chemical caused cancer in laboratory animals. This is related to a *zero level of tolerance*, which means that even if you can measure a few molecules of a suspected carcinogenic chemical it should be banned. If this were taken seriously, we would not be allowed to eat most of the healthy vegetables now consumed daily, including those grown organically. This also applies to modern pharmaceutical drugs. Many modern drugs are toxic, some very toxic, but when prescribed in correct dosages and for specific diseases, they can and do save and extend lives. The EPA has established levels of tolerance for many synthetic chemicals suspected of being carcinogens, but not for natural ones found in food and the environment.

Over the past several decades there have been a number of environmental scares involving cancer or the threat of cancer that have been highly misrepresented and widely publicized. Details for all of them are beyond the scope of this book, but we will mention misconceptions related to several of the more famous ones.

Herbicides

An herbicide can be either inorganic, such as salt, which is an excellent plant killer, or organic, which can be either natural or synthetic. Most man-made synthetic herbicides are of recent origin. Herbicides can be either selective or nonselective in the plants they kill. Agent Orange and dioxin are related herbicides designed to kill selected plant growth that have both been linked to environmental toxicity and the threat of cancer.

Dioxin: Dioxin, which is similar to natural plant hormones, was discovered to be a selective herbicide during World War II. It was called 2,4-D, which stands for 2,4-dichlorophenoxyacetic acid, or just plain dioxin. Soon after, many other herbicides (and pesticides) were manufactured that had a great variety of organic structures. There are many misconceptions and beliefs related to the use and effects of these synthetic herbicides on plants, animals, and humans. Dioxin is a component in the chemical TCDD, which may have some TCDF included and which includes what is called Agent Orange. Dioxin is not only an herbicide, it is also a by-product of the production of wood preservatives, the manufacturing of paper, and insecticides. It was used as a component in some plastics. In the late 1970s and early 1980s environmental activists made unsubstantiated charges against dioxin. It was reported that three ounces of dioxin could kill more than a million people. It has also been called the most toxic chemical created by man. These and similar assumptions were spread by the media, but they have not been borne out. One theory asserted that the human body cannot break down dioxin. Scientific examinations have shown otherwise, and as previously mentioned, many natural toxic chemicals are stored in the body at much higher levels.

Dioxin has been tested on many different animals, and it was shown to be most deadly to guinea pigs, but least harmful to hamsters. Lab tests showed that it required 5,000 times the oral dose to kill half of the hamsters being tested, as compared to guinea pigs. Rabbits, mice, and primates were about 200 times less sensitive to oral doses than guinea pigs. For ethical reasons, humans cannot be intentionally exposed to dioxin for testing potency.

Even so, some prisoners volunteered to be tested for the effects dioxin may have on humans. When extremely high doses were placed on their backs they developed a form of acne, but no cancer. Similar doses given to test animals would have killed the animals.

Dioxin is a toxic substance. It is found in nature at very low levels (just a few parts per quintillion). At these levels it is *not* the cause of all the cancers, birth defects, and miscarriages attributed to it. The National Institute for Occupational Safety and Health (NIOSH) headed up a large epidemiology study of workers in plants that made dioxin. They found an increase in cancers in a few workers only when they had been exposed to excessively high levels of the chemical, far higher than any exposure to which residents at Love Canal, New York or Times Beach, Missouri were exposed. These two geographic areas became centerpieces of the toxic-waste issue. The earth-covered waste dump at Love Canal contained about 80 different chemicals, including dioxin. Not all of the 80 chemicals were toxins, but several were. After the scare began, 250 people moved out of their homes (since then, many of these homes have been reoccupied). A similar scenario existed at Times Beach. There was much publicity, fear, and misinformation about both of these toxic areas.

NIOSH found no increase in cancer for those workers in the dioxin plants who were exposed at about the same levels as the general public. The study concluded that dioxin was not proven to have caused the cancer found in workers, that other factors may have caused an increase in cancers found in workers, and that it was possible that the number of cancers that did occur represented a statistical anomaly. They also found that the workers in general were in better health than a similar control group of nonemployees. In 1984 the American Medical Association (AMA) stated that at acceptably low levels, with the exception of acne, dioxin did not cause long-term biological activity that affected humans' cardiovascular system, the central nervous system, liver, kidney, thymus, or immunologic system. In addition, it has not been demonstrated that dioxin causes any short- or long-term reproductive effects in either men or women.

Agent Orange: The Agent Orange issue is somewhat different, both politically and scientifically. During the Vietnam war the Air Force used almost 20 million gallons of different types of herbicides on the Vietnamese countryside to remove plant foliage so the U.S. forces could see the enemy and not be ambushed. Of this amount, about 11 million gallons was the chemical Agent Orange. It was named for the orange stripes painted on the containers of the herbicide. It is a mixture of 2,4,5-T and 2,4-D, and thus is related to dioxin. Most of it was used as a tree defoliant, but some was used to clear river banks of brush so American ships would not be ambushed by

Vietcong guerrillas. About 94% or 95% of the chemical was spread on the leaves of trees, but a small amount affected crops as well.

The first report that massive doses of Agent Orange could cause birth defects in mice was released in 1970. After some early publicity, and during the next few years, several scientific missions led by the American Association for the Advancement of Science (AAAS), and the National Academy of Science (NAS) could find no evidence of harm to humans from the use of Agent Orange. Vietnamese civilians claimed harm had come to their animals and children, and because the United States did not want to be accused of engaging in chemical warfare, the spraying stopped.

Soon after, Agent Orange was linked to several types of medical problems in many returning veterans. These charges led to several important studies on the effects of Agent Orange on humans. In the 1980s a study by the Centers for Disease Control (CDC) stated that the link between Agent Orange and various cancers was too tenuous to prove, but this did not stop the debate. Since 1979, over 200,000 Vietnam veterans have registered with the Veterans Administration (VA) for medical examinations. One of the first studies to determine harm caused by Agent Orange was conducted by using three different groups of people, those who handled the chemicals, veterans who served in Vietnam, and veterans who did not serve in Vietnam. A follow-up study in 1987 showed a median of 12.4 ppt of dioxin in the blood of members of Operation Ranch Hand, the soldiers who did most of the spraying for the Air Force. This is high when compared to an Air Force study that found a median of 4.2 ppt for the same group. This compares with the 5.0-ppt level found in the ground forces who served in Vietnam, and the 5.0 ppt for veterans *who were never in Vietnam.*

Follow-up studies have been conducted and no statistically significant increases in cancer or birth defects have been identified. A statistically significant number of former Vietnam servicemen had a higher rate of diabetes than did the control group, however, on the average they were more overweight than the control group. There is no explanation for the difference in weight and occurrence of diabetes in the servicemen as compared to the control group, but it may or may not be related to Agent Orange. In 1986 a VA study was conducted that involved almost 14,000 patients in VA hospitals who were Vietnam veterans. The study found no difference in the occurrence of cancer between veterans exposed to Agent Orange and those who were not.

About this time there was much publicity that confused animal studies with health conditions of actual veterans. As mentioned, dioxin was shown to cause some cancers and birth defects in a few types of animals. This data was extrapolated to humans. In addition, the animal data was used to indicate that it also caused birth defects in the children of returning veterans.

Several studies were conducted to see if this was the case. One study concluded that data collected contains no evidence to support the position that Vietnam veterans have had a greater risk than other men for fathering babies with all types of serious birth defects combined. Certain subgroups of Vietnam veterans' children had a lower increase in birth defects than did a larger group's children. When the results were averaged for all Vietnam veterans, the studies indicated that there was no increase in birth defects of their children. An unexpected result was found by the same study—that children of Vietnam veterans had about half the expected number of birth defects classed as "complex cardiovascular defects."

Of some interest is the effort of scientists to use genetic engineering to develop organisms that can consume toxic-waste materials. Several decades ago the first patent was granted for an engineered organism that could digest spilled petroleum. Efforts to develop other altered microorganisms that could clean up dioxin and Agent Orange by breaking down the toxins into harmless substances has continued.

Electromagnetic and Nuclear Radiation

There is a general misconception that electric powerlines affect the environment and that harmful radiation results from this source. These fears are the result of not understanding the nature of both electric and magnetic fields. EMF stands for *electromagnetic fields*; it is sometimes referred to as *electromagnetic radiation,* which is not the same as nuclear radiation. They are two related but separate physical phenomena that are often confused. Electric currents have their own type of field, as does magnetism, but a flowing electric current always has a related magnetic field surrounding it (see Figure 6.1, which demonstrates an easy way to determine the direction of

Figure 6.1 Right-hand rule. Grip wire with your right hand with thumb pointing in the direction of the flow of current. The direction of the field will be the same as that formed by your fingers around the wire.

the magnetic fields related to the direction of the current). Radiation exists over a large range of wavelengths and frequencies composing what is call the *electromagnetic spectrum* (see Chapter 4). At the very long wavelength end of the spectrum are electric, radio, and TV waves (EMF radiation) of very low frequencies. Frequency refers to the number of vibrations or cycles per second of the wave. Microwaves, which are similar to radar, are somewhat shorter, then comes the visible-light spectrum starting with the longer infrared wavelengths, followed by the colors of the rainbow, which are followed by the ultraviolet waves of shorter length. As the wavelengths of radiation of the electromagnetic spectrum become shorter, their frequencies become greater. Therefore they have more energy and become more penetrating and harmful. In a sense, all electromagnetic radiation is light. It is just that some wavelengths of radiation are not visible to the unaided eye. But we do have instruments to detect electromagnetic waves that we can't see with our eyes; for example, radios and TVs, radar screens, infrared detectors, X-ray film, Geiger counters for gamma rays, and so on.

There is a very distinct difference as to the harm that can be caused by radiation of different strengths, frequencies, and wavelengths. The radiation of longer, low-frequency wavelengths does not cause ionization, whereas the shorter, high-frequency, more penetrating radiation does cause ionization of atoms and molecules in our cells and tissues. An atom of an element has a neutral charge; that is, it has the same number of negative electrons in its orbits or shells as the number of positive protons in its nucleus. When short, energetic radiation ionizes an atom or the atoms in molecules, cells, or tissues, they either gain or lose electrons. Thus they end up with an electrical charge and no longer have the same chemical or physical characteristics as did the neutral atoms, and may become harmful to plant or animal life. If ionization, and many other characteristics of radiation are not understood, many beliefs and misconceptions about radiation's effects on plant, animal, and human life can result.

Electromagnetic Fields (EMFs)

First of all, the radiation in electric power lines consists of very long wavelengths with very low frequencies, thus as radiation they have no harmful penetrating power. We are not referring to the electric voltage or current itself, but rather to the magnetic field generated by the flow of current through the wire. The long-wave/low-frequency electromagnetic waves surrounding a wire carrying electricity are much less harmful to humans than is ordinary sunlight.

The scare of power lines damaging the health of humans is one of the more recent major misconceptions reported by environmental alarmists. The source of much of the misinformation is Paul Brodeur's *Currents of Death*

(1989). He claims that the most covered-up public-health hazard for Americans is the pervasive nature of electromagnetic fields. The facts prove otherwise. When an electric current is forced to flow through a wire, a magnetic field is generated around the wire. Both types of fields (electric or magnetic) can exist together or separately. In addition, there can be different low-frequency electric fields, just as there can be different levels of magnetic fields. Brodeur's misconception is that extremely low-frequency electric and magnetic fields can be hazardous to human health in several ways (he is not referring to being electrocuted). Brodeur claims that, over the long term, overhead power lines, electrical substations, electric blankets, TV and computer screens or video display terminals and even electrical appliances such as air conditioners, toasters, can openers, refrigerators, and electric razors will cause health hazards. Mr. Brodeur's misinformation has convinced many people to blame EMF for their misfortunes. Some examples of the diseases claimed to be caused by EMF are various cancers (including leukemia), brain tumors (from sleeping too close to the electric alarm clock or using a cellular telephone), depression, nervous-system dysfunctions, birth defects, and spontaneous abortions.

One misconception is that it can be harmful to be anywhere near a power line or any source of flowing electricity. The strength of electromagnetic radiation follows the inverse square law, which means that it loses its strength rapidly over distances, so the farther you are from it the weaker it is. Thus the EMF measured from power lines that are over 25–30 feet above ground are very weak at ground level and have no harmful effect on humans. This is particularly so if you are in a house or school. Dr. Eleanor R. Adair (1989, p. 11) of Yale University and a member of the Institute of Electrical and Electronics Engineers (IEEE) Committee on Man and Radiation, is a critic of what she calls "electrophobia," which she defines as "the irrational fear of electromagnetic fields." She does not feel that the billions of dollars that will be spent by the public for litigation, product redesign, and burying of power lines is justified, as there is no scientific evidence to support the claims that electromagnetic fields cause cancer, birth defects, or miscarriages.

One study begun in the 1970s by Wendy Wertheimer, a psychologist in Boulder, Colorado, groups families in areas of power lines. She found cases of children living in areas of high-current flow to have a higher risk for leukemia. Her methodology and conclusions have been questioned by scientists. Her studies are of an epidemiological (statistical) nature, which implies a built-in bias when subgroups are isolated from the total group. When subjects are self-selected according to their own identification of a health problem, and then studied separately from the total affected household population, there is cause for concern as to the statistical accuracy of the data. Studies similar to Wertheimer's will be refined and continued for several decades.

The misconception is that there has been a steady increase in cases of childhood leukemia as electromagnetic fields have increased over time. There is a general misconception that childhood cancer is a common disease that is on the increase. Actually, childhood leukemia is very rare. There are only about 1,600 cases reported annually. The fact is that the amount of EMF over the past 50 years has increased many times, but the incidence of childhood leukemia has remained constant (Haney, 1997). Dr. Lawrence Fischer of the Institute for Environmental Toxicology at Michigan State University headed the advisory committee for a $4.5 million study for the National Cancer Institute that compared 638 youngsters with lymphoblastic leukemia with 620 healthy children and histories of exposure to electromagnetic fields of their mothers during pregnancy. The study measured magnetic fields in all the houses in which both groups of children lived, as well as the homes where their mothers lived when pregnant. The study found no evidence that magnetic field levels in the homes [of the research subjects] increased the risk for childhood leukemia.

There is some question as to the statistical significance of studies that established some relationship between very powerful EMF at close range, to an increase in the incidences of cancer. It seems that most studies that establish some relationship between EMF and cancer (and other health problems) have neglected to take into consideration such things as active and passive smoking, using alcohol or other drugs, workplace environments, local pollution, lack of exercise, diet, stress, and so forth. Because not much is known about the science of EMF as related to diseases, there is the possibility that, by itself, it does not cause cancer, but rather assists other carcinogens to bring about conditions that might facilitate the incidence of cancer. What happened with the EMF scare is that some environmentalists approached what they saw as a potential problem differently than would most scientists. First, they identified a potential problem—cancer. Then they discussed this perceived problem with scientists and epidemiologists who say there is no scientific evidence for the problem. Then a cover-up is claimed, which is used in writings and public appearances to sell a particular point of view. This procedure may lead some people to believe the alarmists rather than the results of scientifically controlled research studies.

Nuclear Radiation

Excessive exposure to high-frequency, very short-wavelength radiation, such as X rays, gamma rays, and cosmic radiation is not only harmful, but deadly, depending on the strength and length of exposure, and distance from the source. Only 10 micrograms of radioactive fallout deposited directly on the skin is needed to kill a person. It is deadly.

A big question is how much or how little radiation is excessive? After all, we constantly receive natural radiation from radioactive elements in the Earth and space (e.g., uranium, radon gas, and cosmic radiation, etc.). This background radiation must also be taken into consideration when adding up a total dose. One must consider that short-wavelength radiation is not only ionizing, it is also cumulative; that is, it builds up in the body over a lifetime. When exposed to strong gamma or cosmic radiation, neutral atoms and molecules become charged particles called *ions*. This charge can affect the chemistry of the cells in living tissue. Its potential harm to humans can be diminished with shielding and distance from the source. Health concerns about nuclear radiation became more acute with the development of nuclear weapons. (An explosion of atoms involves the energy of the electrons orbiting the nucleus, and is a "chemical" explosion or reaction such as TNT, whereas a nuclear bomb involves the nuclei of atoms rather than the electrons. Therefore it is incorrect to refer to a *nuclear* bomb as an *atom* bomb.) Regular bombs are "chemical" weapons.

The problem begins in peoples' minds; they are not only fearful of nuclear war, but are afraid of all things nuclear, such as the electricity produced by nuclear power plants. A major misconception is that the nuclear bomb and nuclear power plants do the same thing. The bomb explodes through an uncontrolled, self-sustaining chain reaction of a critical mass of nuclear material (10 lb. of plutonium), whereas nuclear power plants, using similar radioactive substances, have built-in fail-safe systems to control the rate of the fission reaction. A chain reaction is not possible if the rate of fission is controlled. The electricity produced by a nuclear power plant is no different than the electricity produced from water flowing through a turbine in a dam, or through burning coal or oil—they all provide the power to turn the dynamos that generate the electricity.

Another misconception is that nuclear power plants are dangerous. In Western Europe and North America no human life has been lost due to radiation from nuclear power stations. This zero-death rate does not exist for the coal and oil plants that generate electric power. The nuclear-power-plant disaster in Chernobyl, Russia was caused by an improperly designed power plant, as well as human error. The accident at Three Mile Island in Pennsylvania released less radiation than people living in Denver, Colorado receive in 1 year. As previously mentioned, because its elevation, Denver has less atmosphere to block radiation.

The problem of what to do with radioactive waste from nuclear power plants needs to be solved. A possible solution would be to find or develop a market for the waste product. Nuclear waste could be used to develop small nuclear generating systems, similar to those currently used in spacecraft, or to provide electricity for general household use. Someday it may be possible to buy a house with its own mini-electric-power station.

Alar, BST/BGH, and Biotechnology

The use of chemicals to raise and preserve food has long been a problem for environmental activists. Recently, advancements in biotechnology, particularly genetic engineering, have reinvigorated alarmists as they fear the possible consequences of this technology. We will consider some beliefs and misconceptions surrounding both these issues.

Alar

Michael Fumento (1993, p. 19) quotes a segment of CBS's *60 Minutes* that aired in 1989 entitled "'A' Is for Apple": "The most potent cancer-causing agent in our food supply is a substance sprayed on apples to keep them on the trees longer and make them look better." Over 50 million people watched and heard this remarkable statement. Many panicked. The Los Angeles public schools removed apples from the cafeterias; apples were removed from supermarkets; frightened parents called their family physicians. The price of apples dropped by more than 40%. The apple growers in Washington State lost over $135 million that year; many filed for bankruptcy.

The chemical at the heart of this controversy is known as alar, or daminozide. It is a growth regulator used to promote uniform ripening of fruit and to improve the apple's appearance. At the time of this broadcast, alar had already been used for over 20 years without ill effects. The general public believed it was being poisoned. To halt the frenzy, the National Academy of Sciences issued a report calling on scientists to stop questionable research practices related to alar; to stop misrepresenting speculation as fact; and not to release research results to the popular press *before* they are evaluated and judged to be valid by fellow scientists. Once a false charge about an environmental issue is made it is seldom possible to rebut it—it becomes fixed in the public's mind.

The Environmental Protection Agency, the U.S. Department of Agriculture, and the National Institutes of Health all conducted risk studies on alar using both animal models and humans. None of these studies indicated a risk to humans from alar used on apples. During the processing of applesauce and apple juice, most of the alar is broken down into other substances. The amount on the surface and inside the apple is at the level of a few parts per trillion—not nearly enough to harm people.

BST/BGH

BST (bovine somatotrophin), also known as bovine growth hormone (BGH), boosts milk production by as much as 20% to 30%, thereby using less feed and reducing the consumer price for milk. A number of environmentalists claimed the hormone to be harmful to humans, although the harmful effects were not specified. Supermarkets have been forced to stop selling milk derived from cows that have been given BGH, even though many studies indicate that there is no risk to humans. After numerous studies the Federal Food and Drug Administration (FDA) concluded that BST/BGH is inactive in humans and is 100% safe. Even so, with the public's misconception of its potential harm, it took the FDA 8 years to approve the use of BST/BGH.

There have been some unforeseen side effects with the use of BGH in cows—not humans. Monsanto, the manufacturer of this recombinant hormone, lists the following possible effect on the animals: bloat, diarrhea, a disease of the feet, fevers, reduced hemoglobin, ovary cysts, low reproductivity with smaller calves, and udder infections. There are also feeding problems as high-protein foods are required, including animal meat rather than grain. This could lead to spongiform encephalopathy, known as mad cow disease, which leads to seizures and death. There is no evidence that this bovine disease leads to a similar disease in humans, called Creutzfeldt-Jakob disease, which also results in seizures and death. The practice of feeding cows meat products has been stopped in most countries.

Biotechnology

Several environmentalists have publicized misconceptions about biotechnology and its potential benefits or risks to humans. Some predict global disaster from biotechnological research and consider it a bigger threat than the nuclear bomb. These concerns are not based on the scientific evidence related to biotechnology. The public is poorly informed and therefore has many more misconceptions about the science and nature of genetics, genetic engineering, and biotechnology than they do about environmental disasters. Most people have heard about DNA, but few can explain what it is and why it is important, with the possible exception that it can be used to determine the guilt or innocence of criminals in some way.

One misconception is that biotechnology is involved in cloning humans or developing monsters, and so forth. Genetic engineering is the technology used to exchange DNA between different simple organisms. An example: A host cell, usually simple bacteria such as *E. coli,* is joined with human insulin to produce a commercial drug for the treatment of diabetes. This has saved diabetics millions of dollars by providing a less-expensive source of insulin. Similar techniques can be used not only to develop new drugs, but to correct some of the genetic disorders that plague people.

Another misconception is that genetic engineering will be misused. Yet it is one of the most regulated sciences in history. There are studies being conducted to explore the possibility of using biotechnology to benefit human gene therapy, which could safeguard us from the more than 3,500 possible diseases and defects that can be inherited. The average human carries a number of recessive disease genes. If an individual's mate has one or more of the same recessive genes, the genes will pair up and the offspring can inherit that disease or a disposition for some abnormality. It may become possible to correct or eliminate these genetic problems in the future. Any danger that exists comes not from learning more about genetic engineering, but rather from the social/political policies that determine how this knowledge will be used.

Many hundreds of years people have actually performed genetic engineering by crossbreeding plants and animals. The mule is a hybrid of a horse and a donkey. Most of our flowers (roses), vegetables (peas and beans), and crops (corn and wheat) are hybrids that have been genetically engineered by plant biologists. All life is the result of nature's bioengineering. There are many future benefits of biotechnology, but there are also some concerns. The most important potential harm to humans is the possible accidental release of manufactured pathogens. The dilemma is to neither cause harm to humans nor hinder scientific progress.

Epilogue

To summarize this book would be to repeat it. Its purpose is to provide an historical reference for a variety of scientific developments, beliefs, and misconceptions through the ages. It's aim is to assist individuals in learning how to assess the many myths, legends, superstitions, beliefs, and misconceptions surrounding the sciences and pseudo-sciences from the ancient past to the present. The ability to think critically and act intelligently is crucial to maintain a civilized society. Philosophers, scientists, psychologists, and mystics can supply you with reasons for why we believe what we believe. The following are a few explanations for why we may believe something to be true when the evidence suggests otherwise:

1. It is *human* to believe in response to fear, hope, and prejudice (willing nature).
2. We are *insecure* in not believing (believing in something, no matter how irrational, is better than believing in nothing).
3. We *fear* not believing (the unknown is more frightening than the known).
4. It is *easier* to believe in the unbelievable (than it is to learn and face the facts).
5. We *trust* other people's beliefs or promises.
6. We can be *naive* and too easily led.
7. We can be *ignorant* or *satisfied* with too little information.
8. We can be *complacent*.

About 100 years ago in *The Will to Believe,* William James ([1897] 1956, p. 8) went into great detail about the morality and rationality of why we believe. He stated, "It is wrong always, everywhere, and for every one, to believe anything upon insufficient evidence." In the first century A.D. Seneca said, "Every man prefers belief to the exercise of judgment" (Peters, 1977, p. 69). More recently, H. L. Mencken said, "The most costly of all follies is to believe passionately in the palpably not true" (Peters, 1977, p. 69). And still more recently, Carl Sagan (1995, pp. 209–10) suggested that we all need to develop a "baloney detecting" kit to distinguish between science and pseudoscience, truth and fiction.

It is possible to arm ourselves from the assaults of those who peddle misinformation and misconceptions not based on verifiable evidence. Following are some tough questions one might ask to help analyze misconceptions and claims made by people:

1. Who is the person making the claim? What is his or her background? Training or speciality? Reputation? Publications?

2. Is the claim, program, research, and so on identified with a particular political party, special-interest group, or popular "movement?"

3. Have the results of the research on which the claim is made been submitted for peer review and/or been published in a refereed scientific journal?

4. Are the claims made in a book publicized by a public-relations firm, a tabloid newspaper, a sensationalized TV program, or a popular magazine?

5. Who funded the research? The publication? Who gets the royalties? Who received how much in payments?

6. Once published, do reputable scientists agree with the claims? If some disagree, do they give valid reasons for doing so?

7. Has the research or study followed acceptable methodologies? Did they use proper sampling techniques? Were appropriate statistical treatments of the data used?

8. "The test of all knowledge is experiment! Experiment is the sole judge of scientific truth" (Feynman, p. 2).

9. How would other scientists go about testing the claim? Is it logical, can the test be repeated? Does it make common sense?

10. Finally, does the group supporting the research demand that "action must be taken now!" to solve a problem identified by the reported research, study, or program? Do they claim a specific outcome, such as the date of a predicted "doomsday." Do they insist that a new social order is required to solve the problem they identified? Do they exhibit great moral indignation when others will not believe them, pay attention to them, or follow their prescriptions for saving the world, etc.?

To sum all this up: *Those who will not question deserve to be lied to!*

Most of the beliefs and misconceptions explored in this book were not of an intentionally false nature. Historically, most scientific misconceptions were based on limited knowledge. When men and women propagate myths and untruths, they often have some ideological or economic motive. Despite the statement by poet Thomas Gray (1716–1771), "Where ignorance is bliss 'tis folly to be wise," it is the author's intent that after reading this book you will consider all claims with a critical eye.

MOTHER EARTH

At first the rains did appear
Then a flood of more and more;
Soon at last, calm and clear
Then came ooze and mud on moor.

In time bright sun did show
And grass and tree did grow;
Seeds to populate and restore
The Earth's harvest ever more.

Man became and did perceive
What wonders change could be;
Up came sod, down came tree
Again t'was rain and flood for thee.

Dikes were built, rivers diverted
Mines dug, and chimneys smoked;
Water, air, and land were fouled
Environmental apocalyptics howled.

But wait, there is more. . . .
Gaia still provides a bio nest;
For healing Earth's polluted mess
As our Mother always has before.

R. E. KREBS

Bibliography

Abelson, Philip. "Uncertainties About Global Warming." *Science* (March 1990).

Adair, Eleanor R., ed. *Microwaves and Thermoregulation.* New York: Academic., 1983.

Anderson, Flavia. *The Ancient Secret, Fire From the Sun.* Wellingborough, UK: Thorsons, 1987.

Ashworth, William. *The Economy of Nations.* Boston: Houghton Mifflin, 1995.

Asimov, Isaac. *Asimov's Biographical Encyclopedia of Science and Technology.* New York: Doubleday, 1964.

_____. *Beginnings: The Story of Origins—Of Mankind, Life, the Earth, the Universe.* New York: Berkeley Books, 1987.

———. *Asimov's Chronology of Science & Discovery.* New York: Harper & Row, 1989.

_____. *Isaac Asimov's Guide to Earth and Space.* New York: Fawcett Crest, 1991.

Asimov, Isaac, and Jason A. Shulman, eds. *Isaac Asimov's Book of Science and Nature Quotations.* New York: Weidenfeld & Nicholson, 1988.

"Asthma and the Air You Breathe." *Consumers Report* (August 1997): 39–41.

Bacon, Francis. *Novum Organum.* 1620. Reprinted as *Physical & Metaphysical Works of Lord Bacon, Including the Advancement of Learning & Novum Organum.* St. Clair Shores, MI: Scholarly Press, 1976.

Bailey, Ronald. *Ecoscam, The False Prophets of Ecological Apocalypse.* New York: St. Martins Press. 1993.

Barlow, Jim. "Industry Embraces Built-in Recycling." *Houston Chronicle,* July 20, 1997.

Barnes-Svarney, Patricia, ed. *The New York Public Library Science Desk Reference.* New York: Macmillan, 1995.

Barrow, John, D. *Theories of Everything: The Quest for Ultimate Explanation.* Oxford, UK: Oxford University Press, 1991.

Bates, William Horatio. *Cure of Imperfect Eyesight by Treatment Without Glasses.* New York: Central Fixation Publishing, 1940.

———. *The Bates Method for Better Eyesight Without Glasses.* New York: Henry Holt, 1981.

Beckmann, Petr. *A History of PI (Π)*. New York: St. Martin's Press, 1971.

Boorstin, Daniel. *The Discoveries: A History of Man's Search to Know His World and Himself*. New York: Vintage Books, 1985.

Bork, Robert H. *The Tempting of America: The Political Seduction of the Law*. New York: Touchstone, 1990.

Boyle, Robert. *The Sceptical Chemist*. 1661. Reprint, Kila, MT: Kessinger Publishing, 1992.

Bragdon, Claude. "Wake Up and Dream," *Outlook* (May 27, 1931).

Brodeur, Paul. *Currents of Death*. New York: Simon & Schuster, 1989.

Bruno, Leonard C. *Landmarks of Science. From the collection of the Library of Congress*. 1987. Reprint, New York: Facts on File, 1990.

Campbell, Norman. *What is Science?* New York: Dover, 1953.

Carson, Rachel. *Silent Spring*. Boston: Houghton Mifflin, 1962, 1987.

Cassirer, Ernst. *Language and Myth*. New York: Dover, 1953.

Columbia University College of Physicians and Surgeons. *Complete Home Medical Guide*. New York: Crown, 1985.

Compton's Interactive Encyclopedia, version 5.1. New York: Compton's Home Library, 1997.

Conrad, Peter and Kern, Rochelle, eds. *The Sociology of Health and Illness, Critical Perspectives*. New York: St. Martin's Press, 1994.

Concise Science Dictionary, 3rd ed. New York: Oxford, 1996.

Cordato, Roy E. "Don't Recycle; Throw it Away! Misinformation is Behind the Push to Minimize Trash." *Valley Morning Star* (Harlingen, TX), February 20, 1996.

————. The Free Market. Auburn, AL: Ludwig von Miles Institute, n.d.

Cotterell, Arthur. *The Macmillan Illustrated Encyclopedia of Myths & Legends*. London: Marshall, 1989.

———. *The Penguin Encyclopedia of Classical Civilizations*. London: Penguin, 1995.

Council on Environmental Quality. *Environmental Trends*. Washington, DC: U.S. Government Printing Office, 1989.

Crombie, A. C. *The History of Science, From Augustine to Galileo*. New York: Dover, 1995.

Cromer, Alan. *Uncommon Sense: The Heretical Nature of Science*. New York: Oxford University Press, 1993.

Crystal, David, ed. *The Cambridge Paperback Encyclopedia*. Avon, UK: Cambridge University Press, 1995.

Darwin, Charles. *On the Origin of Species*. New York: Mentor, 1958.

Dauer, Francis Watanabe. *Critical Thinking: An Introduction to Reasoning*. New York: Barnes & Noble, 1989.

Davis, Paul. *About Time*. New York: Touchstone/Simon & Schuster, 1995.

D'Emilio, Frances. "Pope Claims Science, Faith Can Coexist." *Valley Morning Star* (Harlingen, TX), October 25, 1996.

Derry, T. K., and Trevor I. Williams. *A Short History of Technology: From Earliest Times to A.D. 1900*. New York: Dover, 1960.

Dictionary of Cultural Literacy. Boston: Houghton Mifflin, 1988.

Dobzhansky, Theodosius. *Genetics and the Origin of Species,* 1937. Reprint, New York: Columbia University Press, 1982.

Dodes, John E. "The Mysterious Placebo." *Skeptical Inquirer* 21, no. 1 (Jan./Feb. 1997): 44–45.

Efron, Edith. *The Apocalyptics: How Environmental Politics Controls What We Know About Cancer.* New York: Simon & Schuster, 1984.

Ehrlich, Paul. *The Population Bomb.* New York: Sierra Club/Ballantine, 1968.

Ehrlich, Paul, and Anne Ehrlich. *Healing the Planet: Strategies for Resolving the Environmental Crisis.* New York: Addison-Wesley, 1991.

Ehrlich, Paul, and Anne Ehrlich. *The Population Explosion.* New York: Simon & Schuster, 1990.

Environmental Quality: 21st Annual Report of the Council on Environmental Quality. Washington, DC: U.S. Government Printing Office, 1991.

Erickson, J. D., et al. "Vietnam Veterans' Risks for Fathering Babies with Birth Defects." Journal of the American Medical Association (1994).

Euclid. *Elements of Geometry. The Thirteen Books of Euclid's Elements,* ed. Gail Kay Haines. New York: Dover, 1989.

Ferris, Timothy. *Coming of Age in the Milky Way.* New York: Doubleday, 1988.

Feynman, Richard P. *Six Easy Pieces: Essentials of Physics.* New York: Helix/Addison-Wesley, 1995.

Flaum, Eric. *The Encyclopedia of Mythology.* Philadelphia: Courage Books, 1993.

Ford, Brian, J. *Images of Science: A History of Scientific Illustrations.* New York: Oxford University Press, 1993.

Forrester, Jay. *World Dynamics,* 2d ed. Cambridge, MA: Wright-Allen, 1973.

Freeman, Ira M. *Physics Made Simple* (revised by William J. Durden). New York: A Made Simple Book/Doubleday, 1990.

Fumento, Michael. *Science Under Siege: How the Environmental Misinformation Campaign is Affecting Our Laws, Taxes, and Our Daily Lives.* New York: Quill/Morrow, 1993.

———. "In the Face of Contrary Evidence." *Valley Morning Star* (Harlingen, TX), September 8, 1996.

Gamow, George. *Mr. Tompkins in Paperback.* Reprint, Cambridge, UK: Cambridge University Press, 1994.

Gardner, Martin. *Fads and Fallacies: In the Name of Science.* New York: Dover, 1957.

Gleick, James. *Chaos, Making a New Science.* New York: Penguin, 1987.

Goodheart, Adam. "Mapping the Past," *Civilization* (March/April 1996): 40–47.

Gordon, Robert E. "Help Landowners Save Endangered Species," *Insight* (May 30, 1994).

Gray, Thomas. "Ode on a Distant Prospect of Eton College,"

Grolier Multimedia Encyclopedia, version 8.0. New York: Grolier, 1995.

Hahnemann, Samuel C. *Organon of the Art of Healing.* Philadelphia: Boericke & Tafel, 1896.

Hall, Marie Boas. *The Scientific Renaissance: 1450–1630.* New York: Dover, 1994.

The Handy Science Answer Book. Compiled by the Science and Technology Department of the Carnegie Library of Pittsburgh. Detroit: Visible Ink Press, 1994.

Haney, Daniel Q. "Study Finds No Evidence Power lines Cause Cancer." *Valley Morning Star* (Harlingen, TX), July 3, 1997.

Hawking, Stephen. *A Brief History of Time.* New York: Bantam, 1988.

Herbert, Wray. "Politics of Biology: How the Nature vs. Nurture Debate Shapes Public Policy—and Our View of Ourselves," *U.S. News & World Report* (April 21, 1997).

Hole, Christina, and Edith M. Horsley, eds. *The Encyclopedia of Superstitions.* New York: Barnes & Noble, 1961.

Holmyard, E. J. *Alchemy.* New York: Dover, 1990.

Homer. *The Iliad of Homer,* trans. Richmond Lattimore. Chicago: University of Chicago Press, 1987.

Horgan, John. *The End of Science.* New York: Addison-Wesley, 1996.

Howard, Philip K. *The Death of Common Sense.* New York: Random House, 1994.

Hubbell, Sue. "How Taxonomy Helps Us Make Sense Out of the Natural World," *Smithsonian* (June 1996): 141–51.

Huff, Darrell. *How to Lie With Statistics.* Reprint, New York: Norton, 1993.

Ihde, Aaron. *The Development of Modern Chemistry.* New York: Dover, 1984.

Isaacs, Alan, ed. *A Dictionary of Physics.* New York: Oxford University Press, 1996.

Jaffe, Bernard. *Crucibles: The Story of Chemistry.* New York: Dover, 1976.

James, Peter, and Nick Thorpe. *Ancient Inventions.* New York: Ballantine, 1994.

James, William. *The Will to Believe and Human Immortality.* Reprint, New York: Dover, 1956.

Kellogg, John H. *Rational Hydrotherapy: A Manual of the Physiological and Therapeutic Effects of Hydriatic Procedures, and the Technique of Their Application in the Treatment of Disease.* Philadelphia: Davis, 1992.

Jones, Judy, and William Wilson. *An Incomplete Education.* New York: Ballantine, 1995.

Kohn, Alexandre. *From the Closed World to the Infinite Universe.* Baltimore: Johns Hopkins University Press, 1957.

———. *False Prophets: Fraud and Error in Science and Medicine.* New York: Barnes & Noble, 1988.

Krebs, Robert E. *The History and Use of Our Earth's Chemical Elements.* Westport, CT: Greenwood, 1998.

Krulak, V. H. "Recycled Myth Ready for Landfill," *Valley Morning Star* (Harlingen, TX), May 22, 1993.

Krupp, E. C. *Beyond the Blue Horizon.* New York: Oxford University Press, 1991.

Leeming, David Adams. *The World of Myth.* New York: Oxford University Press, 1990.

Lerner, Rita G., and George L. Trigg. *Encyclopedia of Physics,* 2d ed. New York: VCH Publishers, 1991.

Levi, Primo. *The Periodic Table.* New York: Schocken/Random House, 1984.

Levin, Roger, "Evolution's New Heretics," *Journal of Natural History* (June 1996).

Libavrus, Andraes. *Alchemy.* 1592. N.p.

Macrone, Michael. *By Jove! Brush Up On Your Mythology.* New York: Cader, 1992.

———. *Eureka! What Archimedes Really Meant.* New York: Cader, 1994.

Malthus, Thomas R. *An Essay on the Principle of Population,* ed. Anthony G. Flew. New York: Penguin, 1985.

Manes, Christopher. *Green Rage: Radical Environmentalism and the Unmaking of Civilization.* Boston: Little, Brown, 1990.

Maranto, Robert A. "It's a Good Day For Optimism: How Can We Save the Earth? Become Capitalists, of Course," *Valley Morning Star* (Harlingen, TX), April 22, 1996.

Margotta, Roberto. *The History of Medicine.* London: Reed International Books, 1996.

Margulis, Lynn, and Dorion Sagan. *Slanted Truths: Essays on Gaia, Symbiosis, and Evolution.* New York: Copernicus, 1997.

Marsh, George Perkins. *Man and Nature: Physical Geography as Modified by Human Action,* ed. David Lowenthal. Cambridge, MA: Harvard University Press, 1965.

Meadows, Donella, et al. *The Limits to Growth.* New York: Universe, 1972.

Michaels, Patrick J. Sound and Fury: The Science and Politics of Global Warming. Washington, DC: Cato Institute, 1992.

Motz, Lloyd. *The Story of Physics.* New York: Avon, 1989.

Motz, Lloyd, and Jefferson Hane Weaver. *Conquering Mathematics: From Arithmetic to Calculus.* New York: Plenum, 1991.

National Academy of Sciences. *Policy Implications of Greenhouse Warming: Report of the Adaptation Panel.* Washington, DC: National Academy of Sciences, 1990.

Newton, Isaac. *Philosopiæ Naturalis Principia Mathematics.* 1650. *The Mathematical Papers of Isaac Newton,* ed. D. T. Whiteside. Cambridge, UK: Cambridge University Press, 1967–81.

North, John. *Astronomy and Cosmology.* New York: Norton, 1995.

Oliphant, Margaret. *The Atlas of the Ancient World.* New York: Simon & Schuster, 1992.

Oxlade, Chris, Corinne Stockeley, and Jane Wertheim. *The Usborne Illustrated Dictionary of Physics.* London: Usborne Publishing, 1986.

Pannekoek, A. *A History of Astronomy.* New York: Dover, 1961.

Paulos, John Allen. *Innumeracy, Mathematical Illiteracy and its Consequences.* New York: Vintage, 1990.

———. *A Mathematician Reads the Newspaper.* New York: Basic Books, 1995.

Petras, Kathryn, and Ross Petras. *World Access: The Handbook for Citizens of the Earth.* New York: Simon & Schuster, 1996.

Petroski, Henry. *The Evolution of Useful Things.* New York: Vintage, 1992.

Piltz, Albert. *The Place of Science Teaching in the Elementary School in Overcoming False Beliefs, Misconceptions, Superstitions, and Fear.* Unpublished research monograph, 1946.

Preston, D. "The Lost Man," *New Yorker* (June 16, 1997): 70–81.

Ray, Dixy Lee, with Lou Guzzo. *Environmental Overkill: What Happened to Common Sense?* New York: HarperCollins/Regency Gateway, 1993.

Read, John. *From Alchemy to Chemistry.* Toronto, ON: General Publishing, 1995.

Real Estate Center. "Private Versus Public Lands." Monograph. Texas A&M University, College Station.

Reese, Charley. "What Price Should We be Paying? Our Freedom is the Cost of Fanaticism by Environmentalists," *Valley Morning Star* (Harlingen, TX), March 26, 1997.

_____. "Politicians Pollute the Air: Gore's Clean Air Standards are Based on Junk Science," *Valley Morning Star* (Harlingen, TX), July 26, 1997.

Ridley, Matt. *The Origins of Virtue.* New York: Viking, 1997.

Ridley, Matt, and Bobbi S. Low. "Can Selfishness Save the Environment?" *Atlantic Monthly* (September 1993): 76–86.

Rifkin, Jeremy. *The Emerging Order: God in the Age of Scarcity.* New York: Putnam, 1979.

Rifkin, Jeremy, and Ted Howard. *Who Should Play God?: The Artificial Creation of Life and What It Means for the Future of the Human Race.* New York: Dell, 1977.

Roberts, J. M. *History of the World.* New York: Oxford University Press, 1993.

Ronan, Colin. *Lost Discoveries: The Forgotten Science of the Ancient World.* New York: McGraw-Hill, 1973.

Ross, John F. "Risks: Where do Real Dangers Lie?" *Smithsonian* (Nov./Dec. 1995): 42–53 (Pt. 1); 43–49 (Pt. 2).

Rothman, Tony. *Science a la Mode: Physical Fashions and Fictions.* Princeton, NJ: Princeton University Press, 1989.

Rubenstein, Ed. "Right Data: Who Regulates Them?: Staffing of Federal Regulatory Agencies (Fiscal Years 1970–1997)," *National Review* (July 14, 1997).

Ruchlis, Hy. *How do You Know Its True? The Difference Between Science and Superstition.* Buffalo, NY: Prometheus Books, 1991.

Rudin, Norah. *Dictionary of Modern Biology.* New York: Barron's Educational Series, 1997.

Sagan, Carl. "Comment and Correspondence: The Nuclear Winter Debate," *Foreign Affairs* (Fall 1986).

———. *The Demon-Haunted World: Science as a Candle in the Dark.* New York: Random House, 1995.

———. "Does Truth Matter? Science, Pseudoscience, and Civilization," *Sceptical Inquirer* 20, no. 2 (March/April 1996): 28–33.

Sagan, Carl, and Ann Druyan. *Shadows of Forgotten Ancestors.* New York: Ballantine, 1992.

Samuel, Peter. "Clean Air, Dirty Politics," *National Review* (July 28, 1997): 20–21.

Santayana, George. *Reason in Science.* New York: Dover, 1983.

Sarton, George. *Hellenistic Science and Culture: In the Last Three Centuries B.C.* New York: Dover, 1987.

Schneider, Herman, and Leo Schneider. *The Harper Dictionary of Science in Everyday Language.* New York: Harper & Row, 1988.

Shabecoff, Philip. *A Fierce Green Fire: The American Environmental Movement.* New York: Hill & Wang. 1993.

Shaw, Russell. "Bucking the Tide of Ecological Correctness," *Insight* (May 2, 1994).

Singer, Charles. *A History of Scientific Ideas.* New York: Barnes & Noble, 1966.

Spangenburg, Ray, and Diane K. Moser. *The History of Science: From Ancient Greeks to the Scientific Revolution.* New York: Facts on File, 1993.

_____. *The History of Science: In the Eighteenth Century.* New York: Facts on File, 1993.

_____. *The History of Science: In the Nineteenth Century.* New York: Facts on File, 1994.

_____. *The History of Science: From 1946 to the 1990s.* New York: Facts on File, 1994.

Spencer, Frank. *Piltdown, A Scientific Forgery.* New York: Oxford University Press, 1990.

Stein, Gordon, and Marie MacNee. *Hoaxes! Dupes, Dodges & Other Dastardly Deceptions.* Detroit: Visible Ink, 1995.

Struik, Dirk J. *A Concise History of Mathematics.* New York: Dover, 1987.

Sullivan, Louis H. "The Tall Office Building, Artistically Considered," *Lippincott's Magazine* (March 1896).

Suprynowicz, Vin. "Save a Species: Try Capitalism." *Valley Morning Star* (Harlingen, TX), July 7, 1997.

Thain, M., and M. Hickman. *The Penguin Dictionary of Biology.* Reprint, New York: Penguin, 1995.

Todeschi, Kevin J. *The Encyclopedia of Symbolism.* New York: Perigee, 1995.

Trefil, James. *1001 Things Everyone Should Know About Science.* New York: Doubleday, 1992.

Twombly, Robert. *Louis Sullivan and His Life and Work.* New York: Viking, 1986.

Velikovsky, Immanuel. *Worlds in Collision.* N.p., n.d.

Waldrop, Mitchell, M. *Complexity.* New York: Touchstone, 1992.

Walisiewicz, Marek, ed. *The Way Science Works.* New York: Macmillan, 1995.

Weaver, Warren. *Scene of Change: A Lifetime in American Science.* New York: Scribner's, 1970.

Webster, Charles. *From Paracelsus to Newton: Magic and the Making of Modern Science.* New York: Barnes & Noble, 1982.

Webster's II New Riverside University Dictionary. New York: Riverside Publishing, 1994.

Weiberg, Steven. *Dreams of a Final Theory.* New York: Vintage, 1994.

Whiston, W. *New Theory of the Earth.* 1696. Reprinted as *Astronomical Principles of Religion, Natural & Reveal'd.* London: Lubrecht & Cramer, 1983.

Whitfield, Philip. *From So Simple a Beginning: The Book of Evolution.* New York: Macmillan, 1993.

Wiley, John P., Jr. "Wastewater Problem? Just Plant a Marsh." *Smithsonian* (June 1997): 24–26.

Witham, Larry. "Are Biologists Guilty of Bias?" *Insight* (Jan. 29, 1996): 34.

Yates, Frances A. *Giordano Bruno and the Hermetic Tradition.* Chicago: University of Chicago Press, 1964.

Index of Names

Index of Subjects

About the Author

ROBERT E. KREBS is the retired Associate Dean for Research at the University of Illinois Health Sciences Center. He is the author of *The History and Use of Our Earth's Chemical Elements* (Greenwood Press, 1998).